BUILDING A MAP SKILLS PROGRAM

by Beth S. Atwood
with the editors of LEARNING magazine

LEARNING Handbooks
530 University Avenue
Palo Alto, California 94301

Foreword

The map making and decoding activities found in this handbook will help students become acquainted with the informative and exciting world of maps. Hands-on projects will turn students into investigators as they learn about basic geographic concepts, maps as recording devices and other ways maps can be used.

The purpose of this and other LEARNING Handbooks is to help make teaching and learning more effective, interesting and exciting. Beth S. Atwood, a former teacher, is an education writer and educational materials developer. Her extensive experience has been combined with LEARNING magazine's research facilities and editorial depth to produce this down-to-earth and lively handbook.

EDITOR: Carol B. Whiteley
ILLUSTRATIONS: David Hale, Dennis Ziemienski
COVER: David Hale

EXECUTIVE EDITOR: Roberta Suid
EDITORIAL DIRECTOR: Morton Malkofsky
DESIGN DIRECTOR: Robert G. Bryant

Copyright © 1976, Education Today Company, Inc., 530 University Avenue, Palo Alto, California 94301. World Rights Reserved. No part of this publication may be reproduced by any mechanical, photographic, or electronic process, or in any other form, nor may it be stored in a retrieval system, transmitted, or otherwise copied for public or private use without prior written permission from Education Today Company, Inc.

Library of Congress Number: 76-29234
International Standard Book Number: 0-915092-08-5

Book Code: 012 • First Printing October 1976

Contents

Prologue
GETTING EVERYONE HOOKED ON MAPS   4

Chapter 1
FIRST THE REAL WORLD   9

Chapter 2
SAVING AND SHARING DATA   27

Chapter 3
LINES, COLORS AND LABELS   45

Chapter 4
SOURCES, TOOLS AND KEYS   69

Resources
    89

PROLOGUE

Getting Everyone Hooked On Maps

Maps aren't relevant any more. Photographs have replaced them. For the here-and-now teacher, that view of maps may be true. There is no question that aerial and space photography have been a great boost to the study of large-scale environmental conditions. But photography is a long way from replacing cartography in terms of detail, flexibility, scope, cost and adventure. For example, there is no way one can photograph demographic conditions on Earth in 1938, 1066 or 4000 B.C.—but a good map can show these things. Nor are we able to make photoplots of a river that will be navigationally useful for any length of time. As for adventure, what would *Treasure Island* or the "Bermuda Triangle" be without maps?

Maps are not only practical and relevant, they are fascinating and eye opening. They can offer a lifetime of adventure as well as a library full of data and a laboratory full of problem-solving equipment. Maps can aid in visual learning for kids—maps can be of great use to history and ecology buffs, underachieving readers, adventure seekers, mystery solvers and puzzle posers. They can be helpful in any number of academic and community pursuits. It is the goal of this book to help you help your students understand and appreciate the informative and exciting world of maps.

USING THE BOOK

The structure of this book is intended to dictate neither teaching sequence nor style. You can select the most appropriate starting point for you and your class and move about the activities as you see fit. The breakdown of skills by viewpoint—map maker, map reader, map user—is designed to present you with spoke-like structures on which you may weave a web of mapping activities. Start spinning the web, moving back and forth from spoke to spoke, from reality to fantasy, from the concrete to the abstract. Sprinkle in plenty of curiosity, pondering, investigation and mystery. Build a jewel-bedecked web. Build a sparkling map skills program.

OUTLINE OF SKILLS

I. Understand basic "geographic" conditions and concepts
   A. Investigate local conditions and concepts
      1. Water
      2. Weather/climate
      3. Vegetation
      4. Human activities
      5. Time
   B. Analyze local geographic conditions
      1. Stability
      2. Extremes and irregularities
      3. Patterns and trends
   C. Identify the positive effects of geographic conditions
      1. Provider
      2. Facilitator
      3. Stimulator
   D. Identify the negative effects of geographic conditions on humans
      1. Starver
      2. Barrier
      3. Repressor
   E. Compare local conditions with regional, continental and hemispheric conditions
      1. Likenesses and variations
      2. Major and extreme differences

II. Understand why geographic data is valuable, thus recorded
   A. Determine why collecting geographic data is necessary
      1. Conditions occur irregularly
      2. Conditions change frequently
      3. Identification of a condition requires comparison

B. Determine when geographic data is useful or necessary
  1. When geographic conditions affect safety and/or source of need
     a) Warn of negative condition
     b) Inform about positive condition
  2. When position on Earth is important
     a) Plan human activity
     b) Study unknown territory and conditions
  3. When changes occur (past, present, ongoing)
     a) Anticipate results
     b) Adjust human activity
     c) Delay or halt natural disasters
  4. When humans want to use features or resources
  5. When study of life patterns and/or cultures is important

III. Understand why the map form is used to record and communicate geographic data
  A. Analyze why the map form is an effective record-keeping device
     1. Data can be recorded graphically
     2. Data can be recorded precisely
     3. Data can be stored efficiently
  B. Analyze the map form as a communication device
     1. Map maker reaches appropriate audience
     2. Map conveys data quickly

IV. Understand what a map reader/user needs to know to be successful
  A. Determine what skills are involved in decoding mapped data
     1. Recognition of maps as coded messages
        a) Types of messages/data
        b) Kinds of symbols
     2. Knowledge of the code used for various kinds of data
        a) Standard map symbols
        b) Standard labels, abbreviations, typography
        c) Keyed symbols
  B. Analyze the organization of the mapped data
     1. Systematic/mathematical plotting of data
        a) Position
        b) Space
        c) Other

2. Objective/deliberate selection and plotting of data
      a) Map purpose
      b) Map clarity
C. Analyze when and how maps can be used
   1. As sources of geographic and demographic data
      a) Names, places, conditions
      b) Time, space, position
      c) Quantity and quality
   2. As tools for research and investigation
      a) Relationships
      b) Patterns and trends
      c) Exceptions to patterns
      d) Change
   3. As keys to adventure, exploration, mystery and the future
      a) Re-creations and illustrations
      b) Incentives
      c) Clues
      d) Speculations

THE NORTH-AMERICAN's CALENDAR

And Gentlemen and Ladies DIARY,

BEING AN

ALMANACK,

For the Year of the Christian Æra,

1 7 7 3.

Being the First after BISSEXTILE or LEAP YEAR. And the 13th Year of the Reign of King GEORGE the Third. Calculated for Meridian of BOSTON, in *New-England*.

ECLIPSES.

THERE will be four Eclipses this Year, 1773, in the following Order.

I. The first of the SUN, March 22d, about 12 at Night invisible.
II. The 2d of the MOON, April 7, visible, calculated as follows,

	h.	m.	s.	
Beginning,	2	39	49	⎫
Middle,	4	3	31	⎪
End,	5	27	13	⎬ Apparent Time
Duration,	2	47	24	⎪ in the Morning.
Digits Eclipsed,	8	21	16	⎪
Moon Latt. N. A.		39	56	⎭

A Projection of the Moon's Eclipse.

The Circle *B*, represents the Beginning ; *M* the Middle ; and *E* the End of the Eclipse, &c.

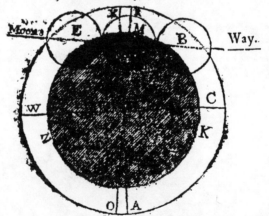

III. The third of the SUN, September 16, a little after Eleven in the Morning. The Limits of the Penumbra will not touch Boston, by Reason of the Moon's South Latitude. Therefore invisible.
IV. The Fourth and last is of the MOON, September 30, between One and Two o'Clock Afternoon, invisible

CHAPTER 1

First The Real World

Reading a map ought to be simpler than reading a written text. Map symbols, frequently pictorial in nature, usually represent tangible, visible and measurable features and conditions. Texts, decidedly abstract in nature, frequently represent extraordinarily complex concepts. What, then, makes map reading and map interpretation so difficult to teach or learn?

Much of the trouble may be blamed on our sophisticated, urbanized society, not on students' inability to decode and interpret symbols or imagine life in distant lands. Most people today simply have limited contact with the basic elements of geography and geographic conditions. Television, photography and other media give them an on-the-spot view of habitats, habits and environments in numerous regions. The life of a pride of African lions is often better known than that of a nearby herd of cattle. The impact of a drought on the Sahel tribes can be more real than the impact of abnormal precipitation on the price of locally grown strawberries. Technology has greatly reduced the need for most individuals to watch the weather; in addition it has eliminated much of the need to pass along knowledge about weather patterns and their implications from one generation to another. What was once common knowledge for many—soil conditions, river flow, as well as weather patterns—now is taught in school science and social studies courses.

Given these circumstances, the problem of poor use and understanding of maps is not entirely inability to decode map symbols representing certain conditions—it also involves the inability to interpret the significance of those conditions. Thus a map skills program is irrelevant if it does not simultaneously develop or refine youngsters' understandings of some specific geographic conditions and concepts.

What specific geographic conditions should be considered?

The geographic "conditions" that have the greatest implications for effective map use and interpretation are: *water, weather/climate, vegetation, man-made features and conditions* and *time*. Land, a passive rather than an active factor, can be studied as an affector of the first four conditions. While this leaves out a large portion of the topics normally classed under geography, a solid understanding of the listed conditions can significantly enhance map use.

What implications do these conditions have on map reading?

The major implications are of continuous change and action of a condition versus an apparently static state portrayed on a map. Even multipurpose maps cannot convey the impermanence of environmental or demographic conditions. A corollary implication is understanding that identifiable patterns and trends are likely to be part of the process of change.

What does a student need to know about these conditions before he or she can become an effective map user?

Students must be able to form mental images of all geographic conditions portrayed on a map before they can make full use of the map. For example, maps that refer to vegetation patterns (taiga, tundra, rain forest) assume the reader has some concept of the basic differences; a map that shows interstate highways, state highways and railroads assumes readers can form a mental image of the real feature. In launching a map skills program, it is probably wise to assume that youngsters' experience with the various conditions is not as advanced as textbook maps imply.

What kind of knowledge is involved in building the basic geographic understanding necessary to relate to maps?

The basic knowledge is physical: recognizing the existence of various features and identifying the characteristics that distinguish one feature from another. On the whole, teachers and texts take time to

present these understandings in as concrete ways as possible. But it is the most complex levels of knowledge that are often overlooked:
- Knowing which physical conditions reoccur often enough to be seen as predictable patterns
- Identifying how geographic conditions affect human (animal) activity both negatively and positively
- Knowing which patterns of human activity (e.g., settlement, agriculture, migration, war) often reflect the influence of geography
- Identifying how time affects the quality and quantity of conditions and the manner in which humans react to them.

How do you include geographic studies in a curriculum already full to overflowing?

You may already be developing more geographic concepts than you realize. Most likely what is missing is a framework with which you can identify areas of misconception and ignorance. The following steps can be used as one framework:

1. Check the science curriculum for the two grades below the one you teach. Identify concepts previously introduced in them that can be reviewed with an eye toward map study. For example, ecological principles of *diversity, adaptation, interrelationship* and *change*; physical and chemical actions of *erosion, oxidation, absorption, reflection, dissolving;* lunar and solar influence on *seasons, day-night cycle* and *weather/climate.*

Check the social studies curriculum of the earlier grades for concrete examples of human activity that is (was) affected positively or negatively by geographic conditions. For example, communities that capitalize on proximity to the sea, fertile soil or extensive grasslands as a source of food and income; inventions that helped humans overcome a severe climate or the space barriers of land and sea.

Check the music, art and language curriculum for examples of the impact of geographic conditions on human culture. These subjects are often overlooked, yet even youngsters divorced from land and sea build geographic concepts from a nation's culture. Include, for example, fiction set in identifiable sections of the country; folk songs describing the efforts to conquer the sea, cross a continent, construct railroads; folk art and fine art that utilize natural resources (wood, stone, gems); folk sayings about the weather or poetry that depict a people's view of the sea, land or seasons (no desert-dwelling people would have a saying, "rain, rain go away").

2. Compare your lists of basic concepts already introduced with the first portion of the Outline of Skills on page 5. If you wish, you can

organize the concepts within that framework. For example, erosion, oxidation and the other physical/chemical actions could be considered as subtopics under (A) Investigation of local conditions and concepts—water and weather/climate. Art and literature portrayng or discussing geographic features unique to a particular region could fall under (C) Identify the positive effects of geographic conditions —stimulator.

Include in the framework conditions with which your students have had firsthand experience. For example, students living in an agricultural district will probably be alert to the effects of wind, drought, flood, frost and heat on vegetation; suburban youngsters will most likely be conscious of the speed (time) with which growing communities change shape and alter the landscape; urban students may recognize the impact of a city's location on the growth or decline of transportation routes or air pollution. And if your class is "typical," a few students will have moved into the area recently bringing firsthand experience of another region or set of geographic conditions.
3. Survey the concepts discussed in the following pages for areas in which your students need some firsthand experience. Try to see that they get this experience prior to or during the initial stages of your map program. Add the more complex concepts for independent or future all-class investigation when the basics have had a chance to jell.

What skills and tools will students need?

To study local conditions, students need to use some of the following investigation processes: observing, measuring, recording, surveying, experimenting and comparing. Tools for such investigation can range from eyes, ears and hands to cameras, barometers and tape recorders. Most of the local geographic conditions suggested in the topics on pp. 23–25 can be investigated using procedures given in science and social studies tests. The aim is simply that when a student reads *river* on a map, most of the implication of *moving water* will be within his frame of reference.

For nonlocal conditions and investigation beyond the capabilities of youngsters, books, photographs, music, art and human references are appropriate. Remind youngsters that informal sources, such as fiction, TV or movie travelogs or science films or a neighbor's account of her vacation experiences in another country, are valid. Authors of juvenile literature are usually conscientious about accurate settings. Photographs often allow a viewer to see something better than he might in real life. A friend may be able to note effec-

BEAUFORT WIND SCALE

Beaufort Force Number	State of Air	Wind Velocity in Knots
0	calm	0–1
1	light airs	1–3
2	slight breeze	4–6
3	gentle breeze	7–10
4	moderate breeze	11–16
5	fresh breeze	17–21
6	strong breeze	22–27
7	moderate gale	28–33
8	fresh gale	34–40
9	strong gale	41–47
10	whole gale	48–55
11	storm	56–65
12	hurricane	above 65

tively differences between customary vegetation patterns and new ones.

What special terms will students need to know prior to map study?

Terms peculiar to mapping are best learned in the appropriate map context. But a large portion of the words used while working with maps is probably already in a student's vocabulary, perhaps taught and used in other subject areas. A review of the following words is advisable, stressing, however, precise rather than general meaning (the terms are assumed to be at approximately fourth grade level; you can adjust them to the grade level you need): symbol, data, set, grid; height, length, depth, width, distance, scale; observe, collect, survey, measure, match, compare, plot (trace, mark); event, cycle, route, population, communication, transportation, diverse, regular/irregular, usual/unusual, continue/discontinue; grow, increase, expand, extend; decline, decrease, cease.

What should students understand about the five map-related geographic conditions?

WATER: Barrier, pathway, sustainer and modifier

We land-based humans tend to view land as the most significant part of the Earth. In reality, however, three-fourths of the Earth is water, so understanding the properties and roles of water is unquestionably significant to map users.

For a start, students should be familiar with water in its various forms: fresh and salt water, glacier and sea ice; geyser and other subsurface water phenomena (for this book, precipitation falls under the category "Weather/climate"). Students also need to recognize water as a changer or modifier through its capacity (in liquid state) to flow and fill the form of its container (lake, puddle); to dissolve minerals; to carry silt, rock and debris; to erode soil and rocks; (in solid state) to carve out valleys and lakes; to cool and warm nearby land and air.

Students should also know that water is not only a physical modifier, but is a biological and cultural modifier as well. Water is essential for plant and animal life; the presence or lack of water affects plant and animal populations, their diversity and adaptation, migration and other habits.

As a demographic affector, water's roles are staggering. In addition to the impact it has on humans as animals, water serves as:
- A source of food, minerals and power
- Both barrier to and facilitator of trade and travel
- A source of danger, destruction, terror
- A challenge to explorers, seamen, map makers
- A built-in defense system
- An invitation to exploration and exploitation
- A basis for war, great empires, political power
- Inspirations for regional and national achievements in science, art, maritime crafts and inventions

As the list of topics on page 23 suggests, terms related to water are numerous—they can also be confusing. Most map readers find it difficult to distinguish seas from oceans, bays from bights and gulfs. The terms seem to be almost interchangeable, and, in a large sense, they are. Many map terms that apply to water differ only in the words they are derived from—some are Latin-based, others German-based. Students can be alerted to this situation through a study of word origins. They can also collect evidence of water's impact on maps by compiling a list of places whose names involve water or its use, such as Portsmouth, Hartford, Great Falls, Sutter's Mill.

WEATHER/CLIMATE: Predictable and unpredictable

Every region on Earth is subject to the influence of wind, sunlight, atmospheric pressure and precipitation in varying degrees of intensity and regularity. Short-term weather and long-term climate are affected by and affect geographic features and conditions. Many of these conditions occur in patterns that can be plotted on maps or interpreted from maps—providing the reader has a basic understanding of the patterns.

The routine ways for youngsters to familiarize themselves with local weather are by measuring temperature, precipitation, atmospheric pressure, wind speed and direction and amounts of sunlight. However, understanding weather/climate as they relate to maps should go beyond these activities. Students need some knowledge of how the Earth's tilt and revolution affect weather/climate, as well as an understanding of how lunar cycles, land forms and water bodies can affect climate.

The more complex the maps, the more necessary it becomes for students to be alert to probable climatic patterns and their "predictable" impact on physical, biological and demographic conditions. Map makers rely on the map reader to bring general knowledge to maps. Unless conveying data about specific weather or climate conditions, maps do not usually show climatic zones and patterns. It is assumed that a symbol for a vegetation pattern, for example, *prairie*, will convey to the reader both the vegetation type and its related precipitation/temperature pattern. But this only happens if the reader relates vegetation to climate and the specific vegetation pattern with a specific climatic condition.

It is possible to investigate weather/climate without relying on routine activities. Plants, animals, buildings, clothing, local and regional customs, phases of the moon, folk sayings and fiction can reveal an amazing amount of climatic data. The possibilities for

ORIGIN OF GEOGRAPHIC TERMS

Terms listed directly opposite each other are often interchangeable on maps or in place names.

Terms listed directly opposite each other are starred are related to similar geographic features.

Germanic Language	**Greco-Latin Languages**
sea	ocean
bight	bay, gulf
shore	coast
cove	
sound	channel
tide	estuary
fjord (fiord)	
waterfall	cataract
stream	river
mouth*	source*
	meander
	delta
land*	terrace*
island	
hill*	mountain*

meadow, grassland	prairie
flat, low (land)	plain
high	
headland	promontory, cape
cliff*	precipice*
ice, iceberg*	glacier*
geyser	
	volcano
stone	rock
sand	
town	city
road, lane	street
harbor, haven	port
north	
east	orient
south	
west	
	occident
	longitude, meridian
	latitude
	Equator
landmark	

investigation range from measuring tree rings for evidence of wet and dry years to matching building styles to extremes of temperature, precipitation and wind speed, to finding out why weather is important to utility, insurance and airline firms. (See page 23 for additional weather/climate topics.)

The influence of climate on human activity is perhaps the most relevant study for map readers. Understanding how climate has affected patterns of human settlement and agriculture can help map readers "read" climate data that *isn't* plotted. Understanding how drought, flood, freak frost or hailstorm affects plant and animal life can help a map reader "read" demographic data—widespread starvation, mass migration, increased urban population—from maps reporting catastrophic weather. Understanding that in the southern hemisphere seasons are reversed from the northern pattern helps a person reading a map comprehend why one might best try to cross the Andes in January. And understanding the predictable demographic patterns in relation to climate can make the discovery of ice-free ports in northern Norway especially intriguing.

VEGETATION: Man's timekeeper, weatherman and storekeeper

Vegetation patterns are interrelated with soil patterns, land forms

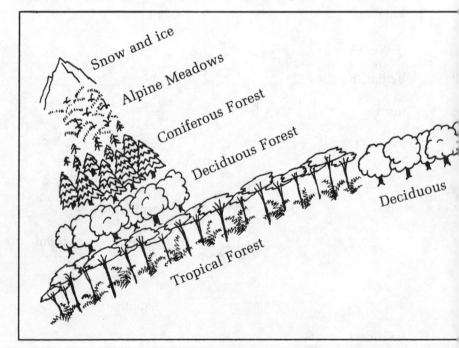

and climatic patterns—and the life cycles and habits of animals (including man) are dependent upon vegetation. Densities of animal and/or human population, local perceptions of (growing) seasons, the cultural value of trees, flowers and lawns, air quality, topsoil conditions and native building materials all reflect vegetation. And since much of this data can be and is mapped, students with a solid understanding of vegetation cycles and patterns can make inferences about vegetation without that specific data being plotted.

In order to make sound inferences, students need to understand how the following factors affect vegetation: sunlight, precipitation, soil type, wind, temperature range, elevation, latitude, land form and relief, natural and man-made disasters. As with climate, the possibilities for investigation of these factors are varied and can be conducted by studying vegetation itself or through investigation of such human-related topics as building and art materials, routes of nomadic herders (grazing animals need vegetation), efforts to overcome vegetational barriers and examples of areas where human carelessness with vegetation has caused catastrophes, changing both animal and human life patterns.

Students should have at least some understanding of the sequence of vegetation types up mountain sides and the sequence of types

Forest   Coniferous Forest   Tundra Meadows

VEGETATION BANDS

from equator to pole. Youngsters also need to recognize the relationship of these vegetation patterns with precipitation, for example, that the miniature plants of the tundra reflect not only polar extremes of light and temperature but also limited precipitation. On a smaller, more local, scale, students could investigate the influence of prevailing winds on trees and man's use of trees and other plants to modify air pollution, erosion, extremes in temperature and featureless landscapes. Then, when faced with the task of comparing maps showing the same area at different time periods, students would be able to "read" with understanding the presence of shelter belts or parks, prairie fires or lumber mills, date palm-dotted oases or beach plum-covered dunes.

HUMANS: Users, changers and circumventers

Human activity has become as important a factor in geographic conditions as vegetation, climate and water; however, its importance is a relatively recent phenomenon. One hundred fifty years ago humans still had to cope with the other geographic conditions on an unequal footing. It is this pre-technological stage that seems most appropriate for beginning map study. Map readers need to perceive of geographic conditions as assets or liabilities to human existence before proceeding with the perception of humans as serious changers of the environment.

Man-made structures are important on many maps, but it is likely that students will need little additional experience learning to distinguish between houses and garages, mills and forts, roads and bridges. It is also likely that students will need to read such structures with a greater understanding of what they reveal about their physical environment. In an age when super machinery and blasting techniques make it possible to bridge almost any river, tunnel through huge mountains and fill deep ravines, youngsters are apt not to see that building could be restricted by land form, water bodies, climatic conditions or vegetation patterns. Yet in much of the world these restrictions still hold true, and historical maps are all but guaranteed to reflect human activity hampered by geography.

To point up this relationship, encourage students to find examples of local conditions where human activity could not overcome geographic conditions, e.g., ground too wet for farming or buildings, areas too rocky or steep for straight roads or airports, coastlines too rocky or regular for safe harbors, climate too severe for certain crops. Then have students investigate the various ways man has managed to circumvent physical barriers and utilize geographic factors. Again, the possibilities seem endless and suitable for all ages and interests. Utilization of forest and ocean products or wind and water power, reliance on island, desert and mountain fortresses, or perennial floods occur in a variety of regions and time periods. The fight to reclaim land from the sea, cross oceans safely and swiftly, turn deserts into gardens and cold regions into comfortable home sites is as old as the Egyptians and as new as the SST.

TIME: Seen and unseen

Except when specifically plotted or noted in titles or printing dates, time is invisible on maps. Yet it is an ongoing element of map reading and interpretation; though maps give the impression of being timeless, the effects of time on the plotted geographic and demographic data is significant. Students need to understand that time cannot be divorced from maps and map use. Students also need to recognize that time is a relative as well as a measurable concept.

The first step toward such realizations is getting the students to measure time without digital watches and TV programs. Long before timepieces, light, temperature, precipitation, vegetation cycles, phases of the moon, positions of the stars and animal migration were man's timekeepers. In becoming reacquainted with these regular, though not always precise, occurrences, students can gain a more natural—a "geographic"—view of time.

Students also need to recognize the newness of standard time and time zones and to consider why it was relatively unimportant that each town kept (sun) time for itself. Also, if students realize how slow travel is by foot or by horseback, the passage of time as portrayed on maps of historical events—journeys, battles, settlements—will seem less plodding.

Is there any way you can evaluate students' general knowledge about geographic conditions as they affect human activity, short of a test?

There is at least one activity you can use to make a general assessment of students' awareness or growing awareness of the understanding they will need for interpretive map reading: simulate the "perfect" island for human settlement, then destroy that perfection via negative geographic conditions and ask the students to note the changes and try to solve the problems. Post the following list and then set up the simulation described as "Paradise Island—Lost."

Geographic Conditions

Positive	Negative
Provides basic needs and safety	Provides less than adequate amounts of basic needs or safety
Facilitates communication, trade and other human activity	Hinders communication, trade and other human activity
Acts as cultural and intellectual stimulus	Limits or destroys energy for physical and intellectual activity

"PARADISE ISLAND—LOST"
Divide the class into two teams. Grant each team ownership of a "perfect" island for human settlement. Suspending reality for a moment, assume that the identical islands are a mile apart in the balmy South Pacific and have every imaginable asset: comfortable, sunny days, gentle rain some time each night, gentle, cooling northwest breezes, no severe storms, fertile soil, a large spring-fed lake with a river running out of it eastward to the sea, a sandy coast with two sheltered bays, tropical fruits and vegetables, grass for a few cattle and goats, a forest and a small but rich opal mine that furnishes enough gems to sell to traders for manufactured goods and items, such as salt and spices, that are not available locally. Each team may determine where the main town and three or four smaller villages—for the zero-population-growth inhabitants—are located and how, or if, roads will connect the towns. (Pictures or rough maps can be made for reference as the project continues.)

Now that the islands are "perfect," Fate (teacher or child) alters them. For example, a volcano may erupt in a different area on each island—what was on the sites is destroyed (without loss of human or animal life, if you wish). The perfect geographical conditions are altered both by the initial disaster and the ongoing presence of a rather high volcanic mountain.

Ask each team to list what is lost, find ways to make up for the loss, determine how the volcano will affect weather patterns, crops, construction, need for new or different manufactured goods, mining or transportation of the opals, etc. Have the teams compare lists, checking to see if each has considered the major problems and if exchange of portable goods between the islands would be useful.

The process can then be continued and the islands altered with other disasters or blessings, e.g., the wind patterns shift, another type of gemstone is discovered, the nightly rainfall increases or decreases, a storm destroys one of the harbors, a boatload of 200 people arrives to live on each island.

Encourage youngsters to discover that even if all the other events occur at the same spot and with the same intensity on both islands, the dissimilar sites of the volcanoes influence every other event, and that the more that occurs, the less alike the islands become both in geographic conditions and in human activity.

Three-dimensional models of the islands and their various features can be constructed for reference in problem solving. Maps showing each new condition can be drawn or plotted later.

TOPICS FOR INVESTIGATION

WATER

Collect local data about:

Number and kinds of water bodies
Water-worn stones, pressure-broken rocks
Where water flows, courses it takes
Old riverbeds, new riverbeds
Water holes and water homes
Warm and cold water
Water serving humans
Water that's too strong, too fast, too salty, too dirty

Identify and label the water body or action:

Fresh and salt water bodies
Waterfalls and rapids
Hidden rivers, crater lakes, geysers
Currents, tides, waves, whirlpools
Shoals, sandbars, reefs
Swamps and deltas
Upstream/downstream, mouth/source
Word history of water terms
Tidal waves and seiches

Research—fact or fiction:

Crater Lake's legendary origin
Paul Bunyan and the Grand Canyon
Loch Ness—deep enough for monsters?
The Great Salt Lake is dead
The Mississippi is growing longer
There's hot water at Yellowstone National Park
Some high tides are dangerous
Little People built Kuapa Pond and Menehune Ditch

Research—influence of water:

What's in the names Portsmouth, Hartford and Grand Rapids?
A culture's attitude toward water— Sir Francis Drake and pirates or Johnny Appleseed and cowboys
A culture's attitude toward water— sea scenes, sea shanties and ships or corn fields, barn dances and prairie schooners
Fantastic farms in river valleys
Water working against the land
The war against the sea
Wars and feuds over water
Periods when power meant merchant fleets and battleships

WEATHER/CLIMATE

Collect local data about:

Daily weather, wind speeds and directions

23

The coldest spot in town
The driest spot in the state
The windiest month of the year
Sheltered spots near school
Clouds as weather forecasters
Atmospheric pressure and wind directions as weather forecasters
Animals and the weather

Identify and label conditions and patterns:

Rainy and dry seasons
Clouds and cloud patterns
Rain, sleet, hail, snow
Summer, winter, fall and spring weather (local, in City X)
Storms we'll never forget
Hurricanes, typhoons, tornadoes
Tropical and arctic air masses
Mountains and precipitation patterns
Regions of extreme precipitation
Hot winds: sirocco, mistral

Research—fact or fiction:

Folk sayings about the weather
Believe-it-or-not records
Greenland was misnamed
It never snows at the South Pole
Not all cultures love sunshine
July is mid-winter in Australia

Research—influence of weather/climate patterns:

Who needs weather forecasts?
Wind patterns around the globe
Winter resorts—warm and cold
What clothes and houses tell about climate
Migrating mammals, birds and butterflies

VEGETATION

Collect local data about:

Where and how does your garden grow?

Native and foreign plants in the neighborhood
Tame and wild grazing animals
Plants that drink gallons of water or almost nothing
What happens to soil without plants?
Telling time with plants
Prevailing wind and plants

Identify and label patterns:

Grasslands (prairie, plains, steppes, savannah, pampas)
Taiga, mixed forest and rain forest
Sequence of vegetation up a mountain side
Matching animal life to vegetation patterns
Deserts, tundra, arctic lands
Subtropical orange groves

Research—fact or fiction:

San Francisco, Rome and Cornwall (England) have palms, but not New York
Many of the world's oldest trees live in North America
Apricots "can't" grow at 57° N. Lat but they do
The world's deserts are growing

Research—influence of vegetation:

Grasslands—zones of movement
Vegetation as barriers to human activity
Human settlement and vegetation patterns
Natural vegetation and agriculture
Reading vegetation patterns with totem poles, masks, cuckoo clocks
Vegetation working for humans

MAN-MADE FEATURES AND CONDITIONS

Collect local data about:

Choosing a location for towns,

farms, homes and factories
How the town grew in shape and size
The most valuable land in town
The cost of water
Mountains cut in half, flattened hills
Stone walls, ditches, dikes, terraces, dams and canals
No one lives there; no one uses that route

Research—humans as changers, users or circumventers:

Cutting back the forests
Fighting back the seas and deserts
Using the forests and seas
Camels, ships and railroads: circumventions of barriers
What a nomad's (farmer's, townsman's) life tells about geographic conditions
Native Americans—careful users of the plains
Life in cold lands and hot lands
Life in the mountains

TIME

Collect local data about:

How long a day is
The shortest day of the year
Using vegetation clocks
From field to forest, lake to swamp
Lunar cycle and the tide
It's winter because I see Orion
What do you mean by "late" ("early," "old")?
Using flood-season and sunlight clocks

Identify and label divisions of time:

Diurnal and nocturnal
Century, millenium, era
Tide means time
The history of calendars
The invention of clocks (and what that meant)
How railroads changed timekeeping

Research—fact or fiction:

Ground-hog Day, the swallows return to Capistrano, the buzzards return to Hinkleyville
Signs of spring, warnings of winter

Research—the perception of time:

Next year is a long way off
The last 75 years were the fastest
The last Ice Age was only yesterday

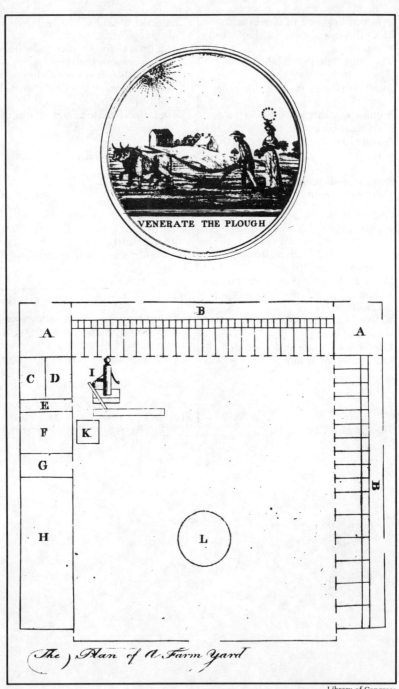

CHAPTER 2

Saving And Sharing Data

Maps are such effective means of communication that almost no effort is spent analyzing why and when mapped data is valuable or why the map form works so well. Yet it is only in investigating these areas that students can grasp the impact of maps as resources.

This chapter focuses on maps as seen through the map maker's eyes—maps as a record of geographic data and a means of communicating that data. Special emphasis is given to helping students recognize the conditions that make collecting and mapping data necessary and to perceive the need for comparative data to describe geographic relationships. While some data plotting in map form is included, most plotting work with mapping symbols and devices is covered in Chapter 3, "Lines, Colors and Labels." The main goal of the following activities is to help students perceive maps as a logical outgrowth of man's need to record and communicate what he has discovered about Earth, be it physical, biologic or demographic.

GOAL: Determine why collecting geographic data is necessary

THEME A: The Earth Is Not a Checkerboard

QUESTION 1: What Features Can't Be Arranged in a Pattern?

Setting the stage: Introduce the term *feature* to your students. Point out that when applied to maps or geographic conditions, *feature* refers to the outstanding or distinct characteristics of a physical, biological or man-made condition, for example, rolling hills, sandy soil, pine trees, bridges, all the space included within the boundaries of Chicago. A map may show water features, land features, vegetational features, climatic features (see Chapter 1). Next, ask teams of students to tour the playground or a nearby outdoor area and list the biological and physical (including man-made) features found there. Encourage them to include such nondemarcated yet known areas as "the fire drill assembly area," features that could be plotted on a map even if their boundary lines exist only in the mind. Review the list with the students when they're ready, and then ask them to compile a master list.

Collecting the data: Have students compile two lists of features that occur in your town (state, country, hemisphere). The first should contain features that humans can arrange in a definite pattern, such as buildings, roads, trees, lawns, farm animals; the second should have features that humans cannot arrange in a pattern, such as rivers, rainfall, mountains, wild birds, oil. Bolder or experienced students can include more abstract arrangeable features, for example, it is theoretically possible to arrange in a pattern much of human activity —provided the geographic conditions are identifiable.

Next, have students find examples in the community or photographs in which biological and man-made features are, in fact, arranged in a pattern—housing developments, trees along streets or walkways, gardens, city streets, telephone poles, street lighting.

Raising questions: Help your students to compare the regularly patterned features with those that cannot be arranged in a pattern, using the following questions: Which kinds of features are possible to arrange in a pattern? What makes the pattern possible? What kinds of features are impossible to arrange in a pattern? What makes a pattern impossible? Which kind of feature is there more of in the world —features suited for patterning or features not suited for patterning? What makes you think so?

QUESTION 2: What Keeps Your Classroom from Being a Checkerboard?

Setting the stage: Provide the students with several checkerboards and Cuisenaire rods that represent 1, 2, 3, and 4 (or similar proportional blocks). Then have the students compile a master list of furniture in the classroom. Ask them to match the furniture size to the block or rod sizes, assuming that the largest pieces of furniture will be represented by the four rods.

Plotting the data: Ask teams of students to make balanced furniture plans of the room on a checkerboard. The goal is to maintain a balance between squares with furniture and those without, and to have no more than two blocks in a square. Since rooms tend to be quadrilateral in shape and furniture movable, this project may seem possible at first glance. However, the odd pieces of furniture guarantee some irregular occurrence of features, e.g., the size, value and single occurrence of a piano make balance or pattern next to impossible.

Raising questions: Focus students' attention on the impact of various features and conditions on classroom geography. Some of the following questions may be useful: Which features (pieces of furniture) occur most often on your floor plan? Why? Which features make a balanced (checkerboard) pattern impossible? Why? Would your floor plan be different at another time of day or year? Why? (There may be more or less features in the room; chairs might be up on desks for cleaning, etc.) If the room were a different shape would a pattern be more possible? Why? If the door or windows were in different places would the pattern be more possible? Why?

Varying the setting: For an alternative to the exercise, use a paved area, such as a parking lot or other level area, that seems "easy" to plot in a checkerboard pattern.

QUESTION 3: In Which Direction Could You Add a Wing to the School?

Setting the stage: Provide the students with measuring tools, drawing tools, paper and, if possible, architectural plans of your school and schoolgrounds, including sites of water pipes, buried storage tanks and the like. Then have the students compute the dimensions of an average room (your room or any other) and use those figures as a guide for a proposed addition to the school. Working together, have the class compose a description of the purpose of the room (or rooms) to be added.

Collecting and plotting the data: Divide the students into teams and ask each team to collect data about the space and features (living and man-made) on one side of the existing school building. Depending on the students' abilities, the dimensions of such features can be estimated by stride or measured down to the centimeter. However, relative positions of features that might hinder construction of the new wing (trees, roads, fences, other buildings) should be accurate enough to determine the feasibility of the project. Ask each team to make a rough sketch of the data it collects and include, if possible, locations of underground features and how the sun travels over the building.

Raising questions: This activity extends the two dimensions of the checkerboard and the classroom (since height was not a factor) so students can begin to perceive that some features occur in more than two dimensions. The activity is also an introduction to the concept of limited space. When all the data have been collected and plotted, ask the students to evaluate the possibility of constructing an addition on each of the sites investigated, using the following questions: Is there enough space for the addition? Will any aboveground features be in the way? Belowground features? Can they be moved? Which site will provide the room with the most sunlight? Shadows? Will that knowledge make any difference in your decision to build? Is it possible to provide more space for the addition? Where and why?

THEME B: A 3-D World Requires Labels and Comparisons

QUESTION 1: How Do You Identify a Place?

Setting the stage: Provide the class with drawing paper and crayons or other coloring tools. Then ask one child to draw and color a shape, design or diagram of a vegetable garden in one section of his paper, keeping the drawing to himself.

Plotting the data: Ask the child to give a step-by-step oral description of his garden to the other students, and have them attempt to duplicate the first child's design without seeing it.

Raising questions: Ask the students to compare their drawing with the original (disregarding the quality of art). How successful were the drawings? Which directions were most helpful? Why? (The most helpful probably gave precise color, size, shape, position.) Which directions were not so helpful? Why? (The speaker may have said "left" instead of "upper lefthand corner.")

Varying the setting: For an alternative to this activity, a child could describe a specific word on a specific page of a book to be located by the class without using any numbers as clues (e.g., a page near the middle of the book has a picture of a black cat in the top right corner; the word is in the last sentence; the word comes after "laugh").

GOAL: Determine when geographic data is helpful or necessary

THEME A: Food, Water and Protection Count

QUESTION 1: Could You Tell Me Where the Nearest Drinking Fountain Is?

Setting the stage: Provide the class with a large map of a zoo or other recreation area that shows positions of drinking fountains, rest rooms or similar visitor conveniences, a large map of a school or town, colored map pins, mural paper, writing paper and art supplies. Then select a topic and specific region for survey that suits the age and research abilities of your students. Point out, by discussion, display or reading assignment, how business and recreation areas serve the public by providing various facilities to make the visit more comfortable. (You may wish to note that these services are less frequent, even non-existent, in some parts of the world.)

Collecting and plotting the data: Assign students the project of surveying a specified region (park, etc.) to learn the location of all the drinking fountains (phones, rest rooms, mailboxes, etc.). When this is done, have the students plot their data on a large map with map tacks, illustrate the data in a mural or on individual drawings/maps or simply list the information under appropriate categories (district, buildings, streets) on a master list.

Raising questions: Help students focus on the concept that data is most likely to be recorded when it is perceived as being valuable to the recorder or to other people. Your discussion may be aided by the following questions: When would local people use the data you've collected? When would a list or map of data be unnecessary to local people? What people would most need your recorded data? Does your data evenly cover the entire area? Why or why not? Is your data useful for every time of day or year? Will it make any difference to your readers if the data is valid at certain times only? Why? How did you identify or describe the location of each drinking fountain (phone, etc.)?

QUESTION 2: When It's Raining Cats and Dogs, Who Cares Where the Water Holes Are?

Setting the stage: Provide your class with a recording or words and music to the song *The Arkansas Traveler*, shoe boxes, modeling clay and other materials for constructing dioramas. Then review the song, stressing in particular the lines describing the man's attitude toward repairing his roof when the sun shone. Ask your students to compare that situation to the situation suggested in the activity's question, i.e., that plenty of rain means knowing where the water holes are is just as unnecessary as fixing a roof that isn't leaking. Encourage your students to think of similar situations or places in which an oversupply of a resource or conditions eliminates the need to record data. For example, a settler in a forested area would probably not bother to record the location of individual trees; a motorist entering a city from a major highway would probably not bother to note the location of each gasoline station.

Plotting the data: Have the students, individually or in pairs, construct back-to-back diaramas. The first diarama should show a setting in which an excess of a positive geographic feature or a lack of a negative factor (e.g., danger) eliminates the need for making special note of its location or existence. The second diarama should show approximately the same setting but, because of a change in geographic condition or human need, there is now a valid reason for recording the location or existence of a special situation. (Ask the students to try to make their diaramas with the questions in the "Raising questions" section in mind.) For example, the first diarama might show a landscape in which a rolling but relatively level land is covered with grass. Here and there are tree-ringed ponds. Travel by foot, horse or wagon is easy; no pathways or guideposts are needed. The second diamara could reflect a different time of year in the same location. Snowdrifts cover the land and ponds. Traveling without guidance and water would be unwise. Other contrasts might be different travel methods, destruction caused by natural disaster, passage hindered by human activity (outlaws, farming).

Raising questions: In your discussion of the activity, try to include the following questions: What conditions in Diarama 1 were plentiful enough to be overlooked? What changed the viewer's outlook in Diarama 2? Could the problem in 2 have been avoided if the features in 1 had been better identified? (For example, the trees by the pond, now covered with snow, were the only clue to the pond's location.) Could the problem in 2 have been predicted? Why? If the problem was unpredictable, would a record of the features shown in 1 make any difference? Why?

THEME B: Sometimes Position Is Everything

QUESTION 1: How Can You Tell If You're Moving Forward If You Can't Tell Where You Started or Where You're Going?

Setting the stage: Provide the class with examples of mazes and dot-to-dot puzzles, a blindfold, a large picture or cutout of a donkey and paper tails to pin on it, a copy of the Greek tale in which Theseus fights the Minotaur in the Labyrinth or a version of *Hansel and Gretel* in which the children leave a trail of crumbs as a guide to their home. Then organize a game of "Pin the Tail on the Donkey" or discuss the game, reminding the students how hard it is to find the donkey once

one has been spun around and lost knowledge of his original position. In addition to the game, have your students try to solve mazes in which the center or end is hidden from view or dot-to-dot puzzles in which most of the beginning and ending numerals, plus others, have been removed.

Collecting the data: Ask your students to collect pictures of geographic settings in which landmarks do not exist (open ocean, underwater, flat, treeless regions) or are so numerous or similar that they are useless as guides (dense forests, desert sand dunes, stalactite-covered caves). Then have the students display the pictures, labeling any problems that a traveler in that area would have determining whether he was lost, going in circles or headed toward his destination.

Raising questions, solving problems: Read aloud or provide for independent reading the tale of Theseus finding his way in and out of the Labyrinth—he used a "lifeline" thread to the outside, in effect identifying his starting point. (You can also use the story of Hansel and Gretel, who try a similar device but less successfully.) In a discussion, try to bring up other situations in which a lifeline has been used as a guide, perhaps during a search for missing cave explorers or hardhat divers. Also encourage your students to solve the problems caused by useless or non-existent landmarks with devices other than

33

lifelines. (*Direction* and *distance* are the most universally practical devices; students, however, may suggest other devices before discovering these means of identifying position.) You can use the following questions for starting your discussion: How can you identify the place you started from without a lifeline? What could keep you from going in circles? Would the time of day affect your travel? Why? Is there a way to find your destination even if it is far from sight when you start? How would you tell another person to get from his starting point to his destination?

QUESTION 2: Will We Need to Build an Earthquake-Proof School?

Setting the stage: Provide the students with a globe, modeling clay, tiny colored flags and one slightly larger flag, a diagram or map of the "Ring of Fire" volcano and earthquake belt around the Pacific or a list of states in the potential paths of tornadoes, hurricanes and high tides.
Plotting the data: Ask a few students to plot the "Ring of Fire" (tornado alley, hurricane paths) on the globe using modeling clay to attach the small flags to areas of potential danger or recent disasters. Use a larger flag to mark the approximate location of your town.
Raising questions: Help students discover that knowing the global position of potential danger areas is important in planning human activity. The following questions should prove useful in a discussion: Where is your town in relation to the earthquake belt? How might that position affect the construction of buildings (shelters, sea walls, utility lines)? Does a disaster happen often enough to keep people from living in a danger area? How can you tell?
Varying the setting: Primary school children may get more from this activity if they focus on local potential danger areas in relation to flying kites, jumping from unknown heights or moving across unfamiliar spaces. Students can measure the extent of their abilities or kite strings and find areas, that is, positions, that are not dangerous.

THEME C: Spotting Change Is Half the Battle

QUESTION 1: Would It Be Wise to Find Another Hiding Place?

Setting the stage: Provide the class with oaktag or file cards, felt tip pens, photographs of scenery or space for a three-dimensional display, miniature buildings, trees, rocks, etc. Then ask the students to build a scene that offers ample place for each student to bury an imaginary treasure. Have each player secretly choose a hiding place, identify it on a file card and put the card in an envelope to be stored with other cards for future reference.

Plotting the data: At different intervals, write on the chalkboard or post a card describing a change in the features of the scene. For example, cards could read: "After a heavy rain, the river rises and overflows its banks 2 feet (2 inches in the scene)"; "The river rises another foot"; "The large pine by the church is cut down to make room for a driveway." Ask the students to note the changes, adjust the conditions in the scene and then decide secretly if their hiding place is in jeopardy. If it is, each student should move his hiding place by noting a new hiding place on his file card. For example, a treasure "buried" near the flooding river could be "moved" away from the river and duly noted on the owner's card. However, if the change destroys or reveals a hiding place, the student should report this fact and withdraw from play. After a predetermined number of changes, have each student reveal whether he still has buried treasure.

Raising questions: In a discussion, try to focus on the following questions: Why did you choose your particular hiding place? Was it a wise choice? If you moved your treasure, was your decision wise? What happened to bring about that decision? Which kinds of changes were easiest to predict? If you had to draw a map showing where your treasure was originally hidden, would it still be useful? Why or why not?

Varying the setting: Students with more experience with maps and/ or geographic conditions could devise a setting showing the construction of a new town, pipe line or highway. They could use "Catastrophe Cards" or "Economics Cards" to show different ways of rerouting and redesigning plans.

QUESTION 2: Do I Have to Adjust the Baking Time?

Setting the stage: Have the students collect and display a few cake mix boxes that specify baking times and directions for cooking at higher elevations. Provide the class with topographical maps, guide books and other reference books that list elevations of a number of towns in your state, if possible, or in states with a variety of elevations, such as Arizona, New Mexico, California, Hawaii, Texas, Alaska and Utah, map pins and a simple map of the area under study. Explain or have students investigate the effect of elevation on the baking process (carbon dioxide is released faster at higher elevations).

Collecting and plotting the data: Ask your students to collect data for several locations in the region under study and list the towns in order of elevation. Then ask the students to plot on a map the location of the towns, using one color pin to represent elevations over 3500, another color for elevations under 3500.

Raising questions: Devise a questionnaire that requires students to read both the lists of names and the plotted data, as well as focuses attention on the effect that different locations might have on baking. Include some of the following questions: In what town will you need to adjust baking time to account for elevation? Of the towns not marked with pins, which ones are most likely to be over 3500 feet? (One example is towns in the midst of mountains over 5000 feet.) In which region is it unlikely that adjustments for baking will be necessary? (Students may say towns along the coast, towns near the mouth of a river.) Which way of showing a town's elevation—rank order, list or map—is the easiest way to find towns over 3500 feet? What makes you think so? (Here the answer depends on a number of factors —total locations shown, number of locations over 3500 feet, distribution of higher elevation sites—so accept any answer the student can defend.)

GOAL: Analyze why the map form is an effective record-keeping device

THEME A: Helping You See Who, What, When, Where and Why

QUESTION 1: Isn't It Easier to Remember a Picture than Written Directions?

Setting the stage: Arrange to take your class to an area of diverse landmarks and features (a park, conservation site, campground, field, beach) where teams of students may plot trails. Provide paper for rough maps and written descriptions, square plastic price tabs from bread bags and small gummed labels for price tabs. If possible, each team should design a trail that does not rely entirely on existing paths. As they work, have them hide several numbered tabs at a specified number of points along their trail. When the trail plan is completed, ask each team to write accurate directions for following the trails, and draw an accurate map of it. Both instructions should include the numbered checkpoints and hiding places of numbered tabs and give the approximate time needed to complete the trail.

Using the plotted data: Divide each team into smaller groups or pairs and give each a written or mapped description of another team's trail. Set the groups off at different intervals or allow one to complete the trail before a second group begins. Tell each group to find the tab at each checkpoint and complete the trail as quickly as possible.

Raising questions: In a discussion, try to include the following questions: Which teams completed a trail first? Why? (Hopefully the map was the most effective guide because of its pictorial aspects.) What was most difficult about using the written directions? What was most difficult about using the map?

Varying the setting: If students are adept at following maps and written directions, the activity may be made more difficult by allowing teams to study the directions or map at the starting point and then try to follow the trail by memory.

QUESTION 2: Why Is the Globe Round?

Setting the stage: Provide the students with a globe, preferably one showing physical rather than political features and, if possible, one or more space photographs that show the curve of the Earth; depending on students' experience with maps, also provide examples of maps on which symbols or color usage is pictorial—but not three dimensional, and writing paper and pencils.

Classifying the plotted data: The object of this activity is not to delve into the evidence proving the basic roundness of Earth, but to point out that because some geographic features tend to be generally or consistently characterized by *shape* or *color*, the feature is symbolized on a map or globe by those characteristics. For example, the oceans, on the globe and most maps, are shown by blue because that is the color that water usually is. There are some areas on the Earth, however, where a body of water is in fact dirty brown, milk white, yellow or almost red; nevertheless, oceans, lakes and rivers are shown as *blue* on a map using color.

Raising questions: Have your students study the globe and/or maps and, using the following questions as a guide, classify some of the mapped data as related to the real feature by (1) shape or (2) color (accept any answers students can defend): Which man-made features (buildings, railroads, roads, bridges, cities, etc.) look like or somewhat like the real thing? How? Which types of vegetation look like or somewhat like the real thing? How? (Answers involving *color* and *shape* are both possible.) What characteristic(s) of islands, continents and capes looks like the real thing—color or shape? What characteristic(s) of lakes, rivers and ponds looks like the real thing?

QUESTION 3: How Do You Draw Yesterday, All Last Week and Next Year?

Setting the stage: Arrange in your classroom a display of pictures of cars, clothes and events (a Model-T, a mini-skirt, one of the moon

landings) that represents a distinct time period. Then provide the class with paper and crayons. Ask the students, through interviews and research, to identify the time period each picture represents. (This may be a single day, a year or a period of years.) Students might also find and list examples of dating methods used in conversation, such as the summer Jimmy was born, just before we moved to Springdale, the year I broke my arm, the day President Nixon resigned.

Plotting the data: Ask the students, individually or in pairs, to develop ways to represent time with pictures or simple tallying devices —but without using our present calendar system. You may wish to display one or two pictures of devices used in pre-literate societies, but too many examples may limit your students' imagination. If students have difficulty focusing on a specific time period, suggest some of the following: yesterday, tomorrow, Fall, 10 years ago, July 4, 1976, Thanksgiving Day, 2000 A.D.

Raising questions: In a discussion, use some of the following questions to focus on the concept of recording time with symbols: Were you able to develop a way to show time without using numerals? Was it easier to show a specific day or a period of time? Could your system be understood by someone else without giving him an explanation? Why would a map maker want or need to date his map?

QUESTION 4: Could You Photograph the Number of Frost-free Days in Barrow, Alaska?

Setting the stage: Select one of the five map-related geographic conditions in Chapter 1 and provide science books and reference books that provide a plethera of data on the topic (for facts about oceans you could try *1001 Questions Answered About the Oceans and Oceanography,* an almanac and an assortment of student books and magazines on the environment noted in the resource section). Then discuss the use of photography, including aerial and space shots, in recording data about the topic. If students have never used a camera, try to bring one to class and let the kids look at objects through the viewfinder, or actually take photos.

Collecting and classifying data: Ask your students to skim the references you provided to find sets of data that can be recorded by photographs (examples of photos would be a plus) and find other sets of data that cannot be recorded with a camera. Have them compare two jointly compiled lists. They'll learn, for example, that while it is possible to take a photograph on one of the few (17) frost-free days in Barrow, Alaska, and on one of the many (348) frost-free days in Tampa, Florida, or in Pine Bluff, Arkansas (228), you can't photograph a period of time in one photo.

Raising questions: Help students perceive the value of "old-fashioned" maps as a device for recording data that can't be photographed, no matter how complex the camera. In a discussion, use some of these questions to focus the research and evaluation process: What kinds of data can be photographed? (Three-dimensional objects; objects existing within camera range at the same time; events and results of events, but not past events, are possibilities.) Is it possible to photograph only the kind of data you want to without recording other data, i.e., can the camera be selective? (Only heat-sensitive, x-ray and similar film can be selective.) What kind of data about human beings can you capture in a picture? (Some answers may be what people build and where they live, but not data about language, nationality, education, religion, etc.) What kind of data can you photograph about historical events? (You can take pictures of locations, landmarks that still exist and permanent changes, but not the immediate results that events have on people, animals, etc.)

THEME B: Almost Everything Is Measured and Matched

QUESTION 1: If the Room Is 4800 Squares Big, How Big Is My Desk?

Setting the stage: Scale—one of the mathematical characteristics of a sound map—is often difficult to grasp and work with, in large measure because map grids must be handmade. (Graph paper is too fine for most student-made maps.) However, needlepoint canvas is accurately constructed and provides a fine surface for colorful flat maps made to scale. To make best use of this activity, therefore, provide the class with needlepoint canvas made of plastic or thread, with the largest holes available (plastic canvas can come as loose as four or five holes to an inch), and scraps of yarn or package-tying ribbon. Also provide large-eyed tapestry (blunt) needles, an illustration of a needlepoint running stitch and an outlining stitch and a floor plan of a room or a diagram of a garden, playground, baseball field, football field.

Plotting the data: Ask the students to collect diagrams of playing fields or gardens on which dimensions are given or to compile the measurements of a room. Demonstrate to them the way 1 foot (yard, meter) can be compared to 1 inch or X holes on the canvas. Have students, working in small groups, convert their floor plan or diagram dimensions, beginning with the perimeter, to the number of holes on the canvas. For example, a room 12'x16', scaled to $1'=1''$, will convert to 60 holes by 80 holes on a canvas in which five holes equal

1 inch. (If scaling is kept to 1, 10 or 100 equaling 1 inch, the math is not difficult.)

Students should begin plotting their maps by outlining the perimeter, using a simple weaving or running stitch or an outline stitch if they can handle needles easily. Features such as furniture, bases or goal posts can be indicated with cross-stitches or solid squares. (Unlike paper, which makes reversion to vertical views tempting, the canvas form tends to keep symbols horizontal and accurate in relationship to all sides of the perimeter.)

Raising questions: Help students to recognize the value of a grid as a guide to *distance* and *direction,* using the following questions: Which was easier to plot (not stitch) on your map—the perimeter or the features? Why? Is it more or less difficult to keep features in relationship to each other on a grid or blank paper map? Why? Could you show the exact path a runner would take on a map of a baseball or football field? How? Could you use a map of a room to figure out the size of carpet to buy for one-third of the room? How?

QUESTION 2: Does It Matter If St. Paul Is on the Wrong Side?

Setting the stage: Provide the class with file cards (two different colors), a chalk board or large piece of paper and road and physical maps appropriate to students' experience. Then make two simple maps, each showing a road that forks at one point. On one map, show a rectangle labeled "red barn" on the right side of the right fork, just past the fork. Then add to the map other features peculiar to the countryside. On the second map, show the same setting, except place the barn just past the fork on the right side of the left fork.

Now have students compare the maps. Tell them that Map 1 has an error, and ask them to find it. Then pose the following question: When will it make a difference that the map maker has made an error on his map? Students may need to invent routes and trace them on the maps with their fingers in order to spot some of the problems. If so, be sure students try approaching the barn both from the main road and from a branch onto the main road.

Next write the following statement on the board: "The exact location of the red barn is important if. . . ." Under the statement, list two or three instances in which the location is important, and two or three instances in which the barn's location is not so important. For example, the list might include:

      IMPORTANT   Your directions were to take the fork past the barn.

	You were looking for a side road just past the barn.
NOT SO IMPORTANT	You came from the fork onto the main road.
	You turned off the main road before you got to the barn.

Now, draw another pair of maps of a similar situation, and repeat the comparison and listing of examples process. Focus again on the accuracy of data being more important at some times than others. (This is *not* an endorsement of inaccuracy, rather an introduction to the value of accurate mapping.) When students understand the concept, have them develop their own maps and statement lists.

Evaluating the plotted data: Ask pairs of students to invent or research a geographical setting in which inaccurate plotting (on a map) of a landmark or other feature could cause problems. Each pair should try to identify two to four instances when exact position is important and at least one instance when accuracy is not so important. When the lists are complete, ask each pair to draw the correct map on one side of a colored file card and the incorrect map (so noted) on the reverse side. Then, on a file card of another color, have them write "The correct location of . . . is important if: . . ." and complete the phrase with one of their statements (have them put each statement on a separate card).

Raising questions: When all the units of maps-plus-statements cards are complete, you can use them as a basis for a team or individual quiz contest. Place a correct map card where several players (or members from two teams) can see it. Then place the accompanying statement cards face down in a pile. Ask the first player to draw a card, read the statement and by referring to the map (both sides if needed) determine whether the statement is important. If it is, the player should say "Very important"; if it isn't, he should say, "Not so important." For example, an older student might be faced with a map of the upper Mississippi River showing the position of St. Paul and Minneapolis. His statement might read: "The position of St. Paul is important if: your sales territory is bounded on the west by the river and you're making a list of all the towns in your district." If the player answers correctly (as determined by the statement's author), he receives two points. If he doesn't, the next player gets a chance and receives one point for a correct answer. When all the maps have been used, the player with the most points wins.

Varying the setting: The game units can be expanded as more types of accuracies—selection of data, correct names—are studied.

A View of Savanah as it stood the 29th of March 1734.

Historic Urban Plans, Ithaca, New York

CHAPTER 3

Lines, Colors And Labels

Most map study programs begin and end with map reading, that is, the decoding of map symbols, the comparison of decoded data and the making of simple inferences. Without question, successful decoding of maps requires practice. But students need more than that to be truly effective map users. They need to be able to classify mapping devices, map purposes and code forms so that skills learned in one map setting can be readily transferred to any map using the same techniques.

The fact that maps are systematically and mathematically devised makes transfer to skills quite possible. Once a reader understands the role of a map key, it makes little difference if the key shows cocoa- and coffee-growing countries or sites of bear and lion cages—a key is always used the same way, no matter what data are coded. And like the key, the skills of finding direction, using scale, reading latitude/longitude and reading population codes can be used with any data in much the same way. Understanding this fact is what sets the regular map user apart from the occasional map reader.

Most of the activities in this chapter involve the reading of maps, not making or using them in exciting ways. Since this is the case, you may want to balance the activities with other more tactile or physical mapping activities. But the most direct and effective way to facilitate map reading, I feel, is to focus on the symbols, devices and systems used on maps. That does not mean map reading cannot challenge the mind and imagination. The fact that maps speak an international language is exciting in itself. Places, names and facts can be fascinating, too. Just as sports fans can envision much of a game's excitement by reading game statistics and scores, a map reader can envision the streets of a far-away city or join the expedition of an arctic explorer simply by decoding a few lines, labels and patches of color. Famous sights and constant dangers are spelled out for all who can read. And the mysteries and puzzles that can be posed by maps lie waiting to be discovered.

GOAL: Analyze the map form as a communication device

THEME A: Maps Are Meant to be Quick, Easy and Special

QUESTION 1: How Do Maps Catch Your Eye?

Setting the stage: Collect, or have your students collect, full-color treasure, travel and adventure maps, maps of remote places and maps with special features, maps with elaborate type styles and with clean, precise type and lines—in short, maps whose topic, cartography or design are eye catching. Also have available books (fiction as well as factual), magazines, newspapers and brochures in which maps are used. (To strengthen this activity you may wish to review "Isn't It Easier to Remember a Picture than Written Directions?" on p. 36.)

Classifying the data: Most youngsters will recognize and identify a map when they see one. And while they are probably aware of maps as informational devices or messages, viewing maps as messages that appeal visually and emotionally—in much the same way as photographs and illustrations—may be a new experience. Encourage your students to look more closely at the deliberate as well as efficient ways map makers use *color, shape, lines, space, order, patterns, type sizes and styles.* You can create a bulletin board display of maps you collect, classifying and labeling the different features. Or you can ask your students to classify in table form some of the most obvious devices used on each map. Accompanying investigations might focus on the frequent use (in promotional maps) of noncartographic devices such as "realistic" illustrations of exotic sights, rare animals and plants, exciting natural phenomena, hidden treasures, challenging or romantic settings.

Raising questions: In a discussion, try to focus student attention on the way map messages can attract and hold readers. You may wish to use some of the following questions: Which mapping devices immediately attract your eye? Which devices are used most often? (Answers will depend on the purpose and complexity of the map under scrutiny.) How are maps and photographs alike? Different? Why are realistic pictures included on some maps? Which of the maps studied was most beautiful (artistic)?

Varying the setting: Older students may be interested in investigating the use of the word "graph" (graphic, -graphy), derived from the Greek "graphikós," meaning *to draw or write*, as it explains the meaning of such map-related words as geography, cartography, demography, topography, hydrography, as well as graph, photograph, telegraph, seismograph, phonograph and thermography.

QUESTION 2: Who's Supposed to Use the Map?

Setting the stage: To the collection of eye-catching maps gathered for the previous activity, add maps that, because of their complexity or monotony, might not be considered appealing, e.g., topographical maps, black and white city maps showing very symmetrical street patterns. Show your students where the date of the map's data is usually printed—in the key, in the copyright notice of the map or, less appropriately, in the copyright notice of the book the map is contained in.

Classifying the data: Refocus students' observation and classification activities toward the *purpose* of the different maps, usually indicated by the map title, caption or the accompanying text. Unless the title is missing, students should not have to decode map symbols in order to find the main purpose of the maps collected.

Raising questions: The following questions can be used to guide group discussion or independent student investigation of the purpose of maps: What is the map supposed to show? (This can be anything from routes taken by arctic explorers to the state birds of the 50 states to roads and highways in North Dakota to camp sites in Washington.) Who might want to use the information shown on the map? Why? Would the reader care if the map was 10 years old? Why?

Varying the setting: When students have had some experience decoding and interpreting maps, the activity could be expanded to include identification and supplementary uses of the maps, such as using a map of political divisions and physical features to study the history of place names.

GOAL: Determine what skills are used in decoding mapped data

THEME A: Map Makers Speak in Code

QUESTION 1: What Do Map Makers Talk About?

Setting the stage: The experience students have had with maps and the kind of learning style you favor will determine your approach to this activity. In it, students may discover the scope and limit of data that can be mapped, either by surveying a range of maps or by surveying a range of data. Since the latter requires some complex analysis of the limits of mapping, that method may be better suited to older students. If you choose the former method, provide the children with a diversity of maps for reference, in addition to paper or file cards that are needed for either form of the activity.

Identifying and classifying the data: Have your students compile lists of the kinds of data that can be communicated on maps. (Keep the lists for reference.) You can use the following categories to focus investigation on general rather than specific data, or guide students toward making these generalizations: physical features (land and water), climate, vegetation, animal life, position of a feature and its elevation, movement and events (natural and human), man-made features (names of places as well as structures), man-made conditions (political, economic, cultural), imaginary places, events and conditions. The kind of data students might list under "movement and events" could include, for example, routes taken by humans, routes taken by animals, paths of ocean currents, prevailing wind patterns, location of natural disasters, location of historical events; the list should *not* include such things as the route I take to school, the wind speed on July 30, etc.

Raising questions: The following questions can be used in a discussion: Can you make a map with only one bit of data? (Be alert to the misconception that a river, lake or island could be viewed as one piece of data. Plotting any of them would require more than one bit of information.) How do you show (define) *space* on a map? (Plot the limits, boundaries, of the space. Even maps of celestial space indicate in some way the area of universe being shown.) Why must there always be a defined space in order to make a map? What helped you decide if data could be mapped? (Answers may include, You could see it, touch it, count it, measure it; it happened at a certain place; the object or condition exists in a certain space.) What kinds of things will not "fit" on a map? (Accept any condition or idea that does not

MAP WORD PUZZLE

1. Sea
2. Direction toward a pole
3. Port or haven
4. Street
5. Land at mouths of a river
6. Guidelines on a map or globe
7. Season from September to December in Southern Hemisphere
8. Strong or protected settlements
9. A precipice or bluff

Answer Key:
1. ocean, 2. north, 3. harbor, 4. road, 5. delta, 6. grid, 7. spring, 8. forts, 9. cliff

occur or cannot be perceived in a spatial framework—the number four, love and so on.)

Varying the setting: Another way to focus attention on appropriate map subjects is through vocabulary projects. Students can build word puzzles, crossword puzzles and analogies to help them develop a better understanding of map terms. They also can construct an illustrated classroom glossary of geographic terms, referring, if necessary, to an atlas glossary.

QUESTION 2: What Kinds of Code Do Map Makers Use?

Setting the stage: Provide your students with a diversity of maps. Then ask them to collect common examples of pictorial and abstract symbols (codes) that communicate information about real and abstract concepts, e.g., international road signs for phone, airport, restaurant, steep hill; silhouettes on rest room doors; dollar signs, mathematical signs; TV channel symbols; initials; arrows. Encourage the students to compare each symbol with what it represents, noting in particular if the symbol bears any resemblance to the real thing. Ask students to invent new symbols to replace some of the more abstract ones, that is, those with no apparent relationship to the real object.

Identifying the data: Now have your class study simple maps and identify the types of symbols (color, pictures or silhouettes, dots, straight, curved or broken lines) and the kinds of labels (place names, direction words, numerals, abbreviations) that are used on them. For younger children, this may be a sufficient introduction to the diversity of map symbols. Older students may wish to speculate on why certain symbols are used and try inventing new ones. For example, buildings are usually shown as rectangles, probably because most buildings in the Western world are basically rectangular; if buildings were generally circular in our culture, perhaps circles would be symbols for buildings.

Raising questions: In addition to general discussion use, students may wish to use the following questions as starting points for investigation and experimentation in map making: What parts of maps are like pictures? What parts are like (or are) writing? What parts use math concepts? (Answers may include numerals, grid pattern, scale, colors or other symbols that represent quantity.) Did you find any maps that used only one kind of symbol, for example, color? Do you think it is possible to use only lines (pictures) to make a map? (Responses here may depend on how students perceive outlines of land masses or water bodies—as use of picture or as use of line.) If a map maker must use only black and white to color maps, what problems arise?

Varying the setting: Older students may be interested in surveying maps published in other languages—they'll discover that maps speak an international language. It is also eye opening to discover that some countries have different names (not just literal translations) for specific places. For example, the Danes call the North Sea the Western Sea; the Germans call the Baltic Sea the East Sea; the French call the English Channel La Manche.

THEME B: Once You Know the Code, You're In

QUESTION 1: What's the Code?

Setting the stage: Provide your class with file cards, grease pencils, photographs (aerial, if possible) of various geographic and demographic settings that are frequently mapped (city streets, rivers, shorelines, parks) and drawing paper about the size of the photographs. (Keep the photographed scenes simple if this activity is an introduction to map symbols.) Then make several copies of the map symbols used and glue the symbol(s) for each feature to a file card (laminate the cards if possible). In order to define the parameters for the map and to show any internal boundary lines, mark lines on the photographs with a grease pencil. Finally, review the concept of maps as messages that can show natural or man-made features or both. (More complex mapping concepts are developed later.)

Plotting the data: Ask each student or pairs of students to select a photograph, study it and list the features shown. Then have the students find the file cards with symbols representing these features, and, using the symbol cards as guides, map the data shown in the photo. (Remind map makers that maps are drawn as though the maker were directly overhead.) Color may be added as the mapper wishes. Encourage students to title their maps, and, if they like, to use the symbol cards as map keys. When the maps are complete, a display showing each photographed area and its corresponding maps can be assembled. Students can also try to match photos to appropriate maps.

Raising questions: As you discuss the photos and maps, you may want to ask the following questions: Why did you choose the symbols you used? If you used color, why did you choose the color(s) you used? Which symbols were easiest to remember? If the photo you used were taken at another time of year, would you use the same symbols for your map? What makes you think so? Who could use

your map? Why would they use it?

Varying the setting: To increase your students' understanding of symbols, have them practice decoding other simple maps using the symbol cards as guides or keys. Students can also match their symbol cards to existing symbols on maps, furthering their ability to distinguish among similarly shaped symbols.

QUESTION 2: Where Is North on a Map?

Setting the stage: Provide the class with a large globe and a diversity of maps using a compass rose or north arrow as a direction key. Then determine where north is in your classroom and chalk a compass rose on the floor or other flat, horizontal surface.

Even if students are familiar with reading direction, they probably need to review the relationship of direction to global position. Point out the North Pole and the South Pole on the globe. Explain

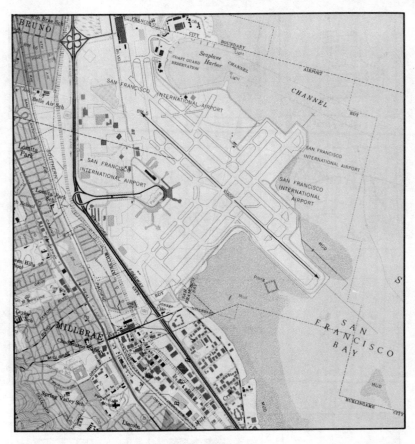

that these points—especially the North Pole—are used to identify all positions and directions on Earth.

Now have students face the North Pole. Using that position, ask them to locate south. If students know how to locate east and west, have them do so. If not, identify these compass points also. Most students will associate east and west with the rising and setting sun. Point out that these are good guidelines, but that the sun's position varies according to the time of year. Show the children examples of a compass rose, north-south arrows and other simplified direction devices (not longitude/latitude). Help older students identify the intermediate compass points *northeast, southeast, southwest, northwest,* and their relationship to the primary compass points.

Identifying and plotting the data: Have your students study various maps and locate the direction device used. Then ask individual students to match the north position on his or her map to north as indi-

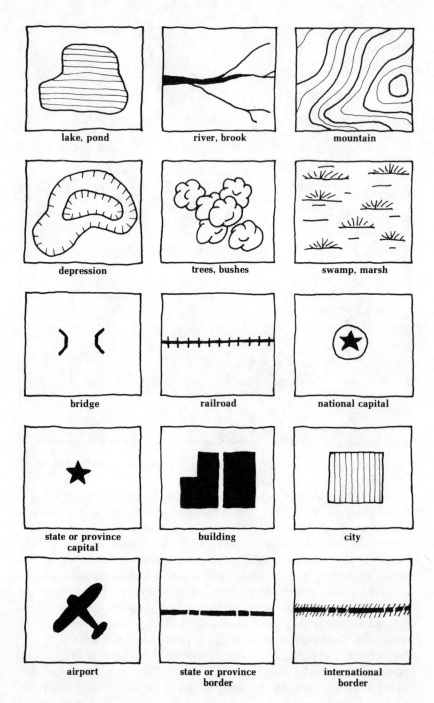

cated by the compass rose chalked on the floor. Be sure the student holds the map horizontally. Emphasize the fact that *north* has nothing to do with the position of the map on the page; north indicates the direction toward the North Pole. If there are no available examples of north in varying positions on a page, search for some, or have students draw some.

Students will probably discover that some maps, particularly continental and hemisphere maps, do not indicate north. Makers of such maps (1) assume that a reader is familiar with the global position of the areas and can therefore locate north without help, and (2) avoid the real problem of indicating that, because of the size of the area shown, north varies from place to place on the map. For example, on a flat map of North America, the straight Alaska-Canada border (approx. 141° W. Long.) runs north and south, slanting somewhat toward the right as it nears the Arctic Circle. On the same map, a line (approx. 80° W. Long) drawn from the eastern bulge of Florida to about the center of St. James Bay, Canada, will slant to the left.

Raising questions: Focus attention on the value of direction being shown on a map by asking the following questions: When would it be essential to have direction shown on a map? (See Question Bl on p. 32.) Why is it important to have all maps use the same direction points? What might happen if north is not shown on a map? (Possible answers are that people might read the map backward or turned sideways.) Encourage students to ask themselves these questions whenever they use a map: Which way is toward the North Pole? Toward the South Pole? In about what area will the sun rise? Set? How would I stand to face north on this map?

Varying the setting: Students may be interested in researching some of these direction-related topics: how early seamen determined direction once land was out of sight; how the phrase *to orient myself* is related to finding location; why Arabic seamen used east as the key direction; why English direction terms are Scandinavian/Germanic (rather than Latin) in origin; how direction on the moon is determined; how celestial direction is determined.

QUESTION 3: How Do You Remember the Code?

Setting the stage: Provide the class with file cards (3" x 5"), sturdy (cardboard, oaktag, masonite) gameboard (24" x 36" minimum), old, duplicate or free maps, felt-tip pens, crayons, glue, buttons or game pieces, dice and a spinner or similar turn-determining device. Then construct or have your students construct one or both of the following map games.

MAP DOMINOES

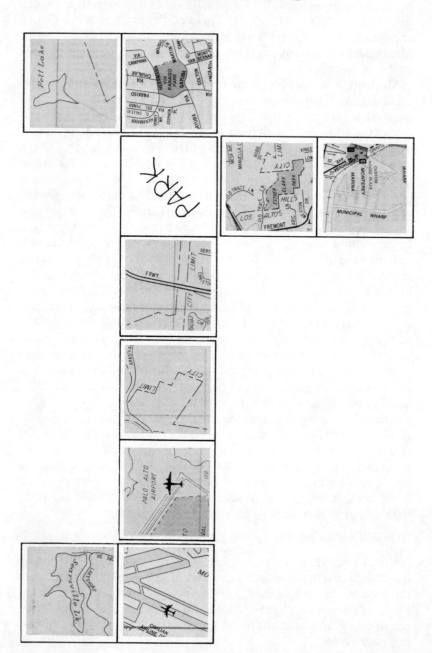

Matching and reading the data: Map Dominoes. Cut small sections from maps and glue these mini-maps to cards domino-style. Be sure that there is one dominant symbol (for a lake, highway, etc.) in each domino half. To extend the number of pieces available for play, terms (e.g., lake, highway, direction) can be written on different ends of a card and can be butted to a mini-map showing the appropriate symbol. Two to four students can play the game, using basic domino rules.

Trip Thru Mapland. On a large gameboard, outline a trip path similar to those used in Parcheesi or Candyland. Divide the path into jump spaces. Then in an irregular but fair pattern, glue a small section of map to a space, alternating maps showing geographic symbols with maps showing man-made/demographic symbols. For example, space 1 could show a river and trees; space 2, some city streets; space 3, an airport area; space 4, a peninsula and water. To play the game, two or four players should decide whether they want to play on man-made symbols or geographic symbols, and then jump their playing pieces, according to the throw of the dice, using the following system: players of man-made symbols move only on odd numbers; players of geographic (physical) symbols move only on even numbers. Each player must keep his type of symbol throughout the game. For example, if player M decides to play the man-made symbols, he moves his piece forward from one man-made symbol jump to the next (no matter how many spaces are in between), but only when he spins or throws an odd number. If both kinds of symbols are shown in the next space he must stop on that space. The winner must throw doubles or spin a predetermined number in order to jump out.

Raising questions: While these games are essentially practice activities, analytical thinking should not be discouraged. If a student can justify a match or move using an overlooked but sound classification of map symbols, then that move should be permitted.

QUESTION 4: How Far Is It? How Big Is It?

Setting the stage: If possible, provide the class with metric linear measuring tools, a watch, strips of oaktag (approx. 2" x 8") and local, state or regional maps on which the scale is shown in kilometers as well as in miles. (NB: Do not use Mercator maps of the world or other projections on which distance is distorted.) Then have students measure out and walk a kilometer (back and forth 10 times over 100 meters will do), four sides of a block or acre or the distance between utility poles. Also have students determine the average time it takes to walk each distance. Since maps often deal in large amounts of space, students may be better able to relate to them if they have a good feel for smaller elements of space, for example, it may help to know that the distance from a student's town to Hartford is the same as 27 times up and down the hill to his house. Students may also use driving or flying time to a specific spot as a reference for mapped distances.

Reading the data: Demonstrate how the distance between two points on a map can be determined using an oaktag measure. Usually it is easier to mark off distance to be measured on the measure and then match that distance against the map scale, but students should be able to compute the distance from the map scale to the measure, too. Also have students practice reading and computing distance using different maps with different scales. If the maps are plotted to show similar spatial quantities (not similar scales), post problems that can be answered using two or more maps. Encourage the use of map reading workbooks and cards published for school use.

Raising questions: As the children work with maps, pose some of these (or similar) work problems: Find out whether town Q or town Z is closer to Lake X. Find three pairs of cities that are about 100 kilometers apart. Find two states that almost match each other in both length and width (Colorado and Wyoming). Find the shortest route to the nearest airport. Find out whether Honolulu or Boston is closer to San Francisco. Find two city blocks (rooms, fields) that are the same length on all four sides.

Varying the setting: For younger students who are not ready to use map scales, relative distance can be determined by using a paper ruler to compare distances (nearer, farthest) between two sets of points or among several points. Older students may find that drawing circles (50 kilometer, 100 kilometer, etc.) centered on a specific city or site will help in the perception of distance. (Polar projections and other azimuthal equidistant projections are constructed on these principles.)

QUESTION 5: How Do You Tell Which City Has More People?

Setting the stage: Provide your students with books that use a diversity of type size, style, face and letter case in chapter heads, subheads, captions, footnotes, etc.; also provide maps that show the population of cities using different dot sizes, type sizes or styles or colors, and paper. Select two maps showing population—one using different dot sizes, one using different type sizes—as demonstration maps. Point out the population key (if shown) and explain how a map

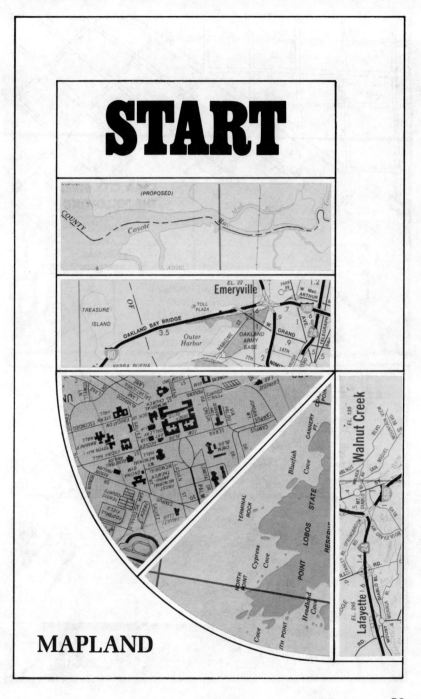

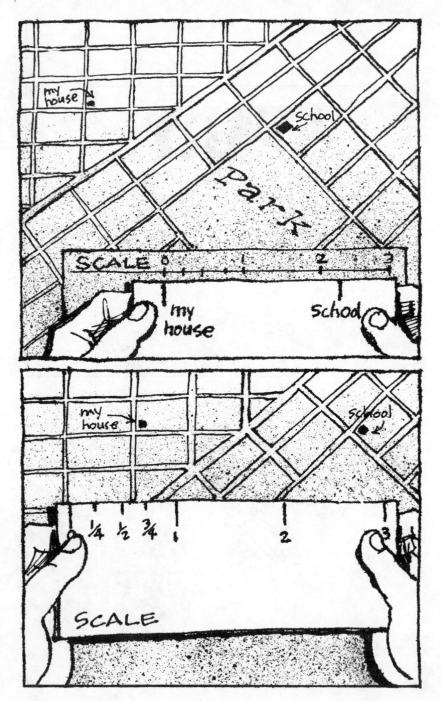

reader uses this code to determine the approximate population of a city.

Reading and classifying the data: In order to do this activity with some degree of understanding, students must be familiar with numbers over 1000 and be able to spot fine distinctions in type size, style and face. If they can do this, have them practice decoding population symbols by picking several size categories and then listing all the cities on a map (or a portion of one) that are a similar size.

Raising questions: Help students to understand map population information by asking the following questions: Is the exact population of each city given? Is it possible to tell which city has the highest population within each category? Why not? In what year were the population figures on your map collected? (This *should* be indicated in the population key.) Do you think that there have been any changes since that time in the size of the cities? Why? How can you check?

Varying the setting: More advanced students may want to investigate maps that show population density. This concept of so many people (animals, trees) per square kilometer (mile) is difficult to imagine, but since students will probably meet such maps in science, social studies and geography texts, it may help them to understand the concept now.

QUESTION 6: What Else Can a Map Key Tell?

Setting the stage: Provide your class with maps that have an assortment of map keys, appropriate in interest and concept to students' mapping experience. (*Washington in Flashmaps* [see the resource section], a paperback atlas, provides over 40 examples of map keys on 46 maps of the nation's capital and its surroundings, and is ideal for classroom map study.) Then have the students locate the key on each map, and tell what kind of information is being coded (restaurants, theaters, concert halls) and what kind of code is being used (colors, letters, numerals).

Reading and reporting the data: Map keys can provide some of the most exciting and colorful data on a map. You can emphasize this point by having your students make three-dimensional map codes, or by having them construct displays in which map data typically shown with symbols are recreated with magazine pictures, yarn arrows and miniature objects such as tiny planes, toothpick TV antennas and modeling clay volcanoes. Product maps can be enlivened with actual product samples and map tacks and flags can help focus on the distribution and quantity of data presented by a map's special code.

Raising questions: Have students compare keys and data on various maps by answering the following questions: How are these symbols like other map symbols? How are they different? (Students should discover that unlike the standard symbols, these symbols are appropriate for only the map on which they appear.) Maps with keys are often called *special purpose maps*—what is the purpose of some maps you've studied? Who might use these maps for this purpose? (A map with the purpose of showing railroad lines in Canada's Atlantic Provinces might be used by a person planning a train trip, a factory planning shipping routes for its products or a vacation home company picking a site for a resort.)

GOAL: Analyze the organization of the mapped data

THEME A: Patterns Keep the Data Precise

QUESTION 1: Where on Earth Are You?

Setting the stage: Provide your students with a globe and hemisphere maps that show the equator, Tropics of Cancer and Capricorn, Antarctic and Arctic Circles and the Greenwich (0°) and 180° meridians, or latitude and longitude lines (grid pattern), modeling clay, black thread and straight pins. Then explain that the lines on the globe are imaginary lines that are used to help locate places on Earth, from north to south and from east to west. (If students are ready for longitude and latitude references, compare this grid to the needlepoint grid used to plot the baseball diamond on page 40.) Post examples of the way positions are written, e.g., 60° N. Lat., 143° W. Long. (the north-south position is given first).

Plotting the data: Ask a small group of students to construct a thread grid system slightly raised from the surface of the globe or map. (Straight pins affixed to the globe with clay can serve as posts on which to tie the threads.) The grid should follow the printed lines under study. Flags should then be placed on cities, islands or other features that fall along the lines.

Raising questions: For many students, the mathematical basis for the various map grids is relatively unimportant. What is important is the ability to locate places using grid references and to use grid references to locate places. Quiz games plus regular use of the grid in map study is probably the least tedious way to develop competency in these skills. For example, students might hold trivia contests to find places whose latitude and longitude designations are approxi-

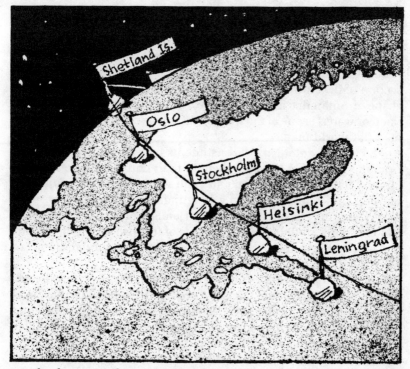

mately the same, for example, Nicosia, Cyprus, 35° N. Lat., 33° E. Long. Or they might pose true-false statements that require careful map reading to determine classification, such as Honolulu is farther south than Miami, Los Angeles, California, is farther east than Carson City, Nevada.

Varying the setting: The time zone map is a variation of the grid pattern. Each of the 24 zones in such a map is basically 15 degrees of longitude wide. (Most deviations are made to accommodate state or national boundaries or to follow natural geographic conditions.) Some students may wish to investigate the history of time zones and the reason for the establishment of a point on Earth for each day "to start."

QUESTION 2: Why Are Maps Different Shapes?

Setting the stage: Provide the class with a globe, tracing paper long enough to circle the globe at the equator, a world map or hemisphere maps shown on several different map projections (polar, Mercator, interrupted, equal area, Eckert) and a geography text or atlas that explains distortions and use of various map projections. Then, using the tracing paper, demonstrate the difficulty of transcribing the entire round surface of the globe to the flat paper surface. Also point out that while latitude lines are parallel and equidistant from each other, longitude lines (meridians) merge at the poles. Have students look at the ways map makers have tried to show the whole or a large section of the Earth at one time. Encourage students to compare the shape of Antarctica as shown on several maps. Point out that shape (land or water) is one of the things that can be distorted or inaccurate on a flat world map.

Analyzing the data: Now have students trace Greenland and Zaïre from the globe and compare them to their counterparts on various kinds of flat maps. Greenland's size and shape are generally distorted on map projections that show longitude equidistant at the polar areas (Mercator) or split the Western Hemisphere through the island (Goode's interrupted). Zaire, by virtue of its position on the equator and toward the center of the Eastern Hemisphere, survives intact on most flat maps and with little distortion.

Raising questions: Help students understand the uses and degrees of accuracy of different maps by asking the following questions: Which maps show scales of miles that are accurate only at the equator? (This should be indicated with the scale.) Which maps would be useless to sea captains? Which maps show north or south in the center of the map? Why do you think the north polar projection is used by pilots flying from the U.S. to Europe and from Europe to northern Asia? (Compare distances on a globe if a polar map is not available.) The Mercator projection shows compass direction in a straight line —why would this be important to ships? Why do map makers accept inaccurate or distorted world maps as a problem that can't be solved? How do they work around the problem? (They use maps with small scales so distortion is negligible, or they use globes.)

THEME B: The Map Maker's Choice Can Hinder or Help

QUESTION 1: What Data Do You See on the Map?

Setting the stage: Provide your students with a diversity of simple maps, a telephone book and writing paper. Also provide, or have students collect, a diversity of tables (single- and multi-column) from almanacs, newspapers, magazines and books on as many topics as possible. Then call students' attention to the phone book and ask what kinds of information (data) they expect to find in the main section and how the data is organized. Point out that the book provides only selected information and therefore has limited use as a reference tool.

Now have each student skim two or three of the collected tables, list the title and the kinds of data given (usually indicated as column titles), how the data is organized and if all or only selected information is given. If the data is selected, encourage students to speculate on how the data was selected. While the reasons for such selectivity and organization may seem perfectly obvious on a table, it may be less obvious, but is certainly important, to maps.

Classifying the data: Have students, working in twos and threes, skim simple maps with an eye to the organization and selectivity of the data that are mapped. The map title and key should help narrow the focus. Depending on the quantity of data shown, students can make lists or multicolumned tables that record the mapped facts.

Raising questions: As they study the map data, have the students answer the following questions: What is the main point of the map? What do you think was the most important kind of data shown? Why? Does the map give symbols for direction and distance? Are distance and direction important to include on your table? Why do you think the map maker chose the data he did? How could you tell it was selected data? (Students might say they compared one map to another or they know from experience that some thing or place is missing.) Does your table(s) have the same visual impact as the map? What is missing? What data are easier to understand using the map? Using the table?

Varying the setting: Students can investigate the impact of too much or too little data shown on a map by plotting data thought to be purposely eliminated by the map maker or data that students think might help make the map more successful. For example, while a map of the area surrounding Kennedy Airport in New York shows many towns, only those familiar with the density of housing in that area have any conception of the number of people affected by airplane noise. Adding a map key and symbols to such a map to indicate population figures for surrounding towns would make the message of the map more striking. On the other hand, including more data on another map might make it too cluttered without adding impact or important information.

QUESTION 2: Is There More Than One Map on the Map You're Using?

Setting the stage: Provide the class with outline maps of various geographic regions, drawing and measuring tools, multicolumn tables

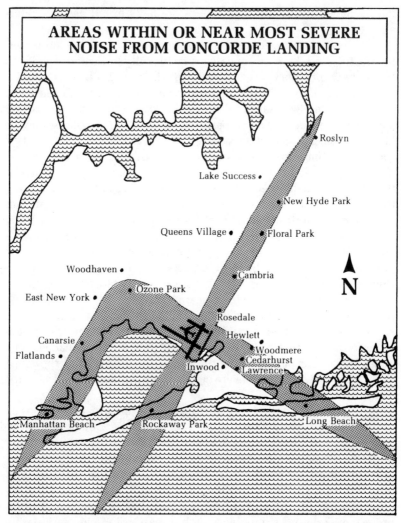

of data that could be plotted on a map and special purpose maps showing more than one set of "special" data. Then ask the students to study a map and select a set of data that could be removed without destroying the base map. Usually this is a set of quantitative data that for simplification of study has been plotted against spatial and physical features to contrast a related set of data, e.g., major crops plotted on a rainfall and physical features map of the Plains states, population distribution plotted on a physical features and vegetation –patterns map of Brazil.

Analyzing and plotting the data: Have the students plot the set of

AREAS WITHIN OR NEAR MOST SEVERE NOISE FROM CONCORDE LANDING		
	Within Flight Path	**Very Near Flight Path**
Manhattan Beach	✓	✓
Flatlands		✓
Canarsie	✓	
East New York		✓
Ozone Park	✓	
Woodhaven		✓
Rockaway Park	✓	
Inwood		✓
Lawrence	✓	
Cedarhurst	✓	
Woodmere	✓	
Long Beach	✓	
Hewlett		✓
Rosedale	✓	
Cambria	✓	
Queens Village		✓
Floral Park	✓	
New Hyde Park	✓	
Lake Success		✓
Roslyn		✓

data they have chosen to remove and another set of data from the same map on separate but identical base maps. Ask the students to use the same key as on the original map.

Raising questions: As they compare the two new maps and the original, have the students try to answer the following questions: Why do you think two sets of important data were put on the same map? Is it more or less difficult to compare different kinds of data on the same map? What kind of data had to appear on each map to make a complete map? (Answers should include physical features, political boundaries, cities, directions.) Why?

CHAPTER 4

Sources, Tools And Keys

Maps are dual-purpose graphics—they are storage places of coded data and effective aids for research, investigation and daydreams. Generally, student and teacher use of maps starts and stops with maps as sources of basic geographic data: What is it? Where is it? How far is it? Which direction is it? What are its neighbors? How does it compare in size, shape or condition to . . . ? This is not an inappropriate use of maps, but much of the data discovered in this way can also be found in reference books.

It is the next level of map use that raises the map from storage place to tool and key. Maps allow scientists, historians and treasure hunters to *see* how the data they are investigating relate to other geographic conditions or data. And these relationships provide incentives for exploration, inspiration for further investigation and clues to mysteries. The activities in this chapter are models for ways maps can be used to help explain, support, stimulate or dissuade investigation of this planet and its inhabitants.

GOAL: Analyze when and how maps can be used

THEME A: Maps Are Sources of Data

Any time maps are used to answer *what, where, how much, how, what kind, who* or *when,* they are being used as sources of data. Since several of the map reading activities and much of the ordinary classroom use of maps fall into this category, no additional activities are included.

THEME B: Maps Are Tools for Research

QUESTION 1: When You Compare Sets of Data Do You Find Related Conditions?

Setting the stage: Provide your class with a large map that (1) shows several sets of data, such as cities, rivers, roads, airports, lakes, and (2) will illustrate a relatively simple example of negative or positive geographic conditions (see Chapter 1 and page 71); miniature planes, buildings, etc., dried or plastic vegetation, yarn, modeling clay, flags and any other devices that can serve as 3-D map symbols, a large sheet of tracing paper if the map is not on coated paper and several world atlases. Then, using the accompanying illustration of the Alaskan map as a model, have your students emphasize several sets of data by affixing 3-D objects to the map. (Modeling clay usually will not stain glossy maps even if the paper is not particularly sturdy.) Remind the students that map makers generally select sets of data for maps, rather than isolated or hit-and-miss items, to fill space.

Seeing relationships between sets of data: Post or make multiple copies of positive and negative conditions that are seen on or appropriate to the map your class is studying. For example, the positive list for the Alaskan map might include ample supplies of game and fish, huge amounts of land, a long coastline with many harbor sites, major rivers running to the sea, ample forest products and valuable minerals. The negative list might include the fact that Alaska's position on Earth fosters severe climate, large mountain ranges hinder travel and settlement, size makes transportation and communication expensive, there is limited land for agricultural purposes, the country's out-of-the-way location hinders cheap or easy transportation of goods to world markets. (NB: This suggested list of conditions ranges from easy to understand to complex; adjust your list to your students' experience.)

Now, have your students flag sets of data that help point out the presence of the negative and positive geographic conditions. For example, the relationship of transportation to Alaska's mountain ranges and great size can be accented by marking the many airports and one rail line with one color of flags, the mountain ranges and land boundaries with another. Isolation from world markets and severity of climate can be emphasized by flagging the overland route

the North Shore oil takes, writing on a plotted flag the days Prudhoe Bay is ice-free and the number of miles from the pipeline terminals to a few major oil customers.

Raising questions: In a discussion, focus on the following questions: Which kinds of relationships were easier to spot—man-made features related to physical conditions or physical conditions related to other physical conditions? Why? (Until students are experienced with the various geographic conditions discussed in Chapter 1, the simpler relationships of man-made features or conditions to physical conditions will probably be easier to identify.) Can you find maps of other areas that show the same kind of relationships? (Several large landlocked African nations and, to a lesser degree, Australia, reflect similar relationships between size, mountains and transportation methods similar to Alaska.) Could you use all kinds of maps to find evidence about a specific relationship? Why? When would it be valuable for a map user to identify this relationship? (See Outline of Skills, II B, 1–4.)

Extending the activity: As students become adept as spotting relationships on maps, compile a poster-size table or file card index of areas that show similar relationships. The possibility of finding parallel geographic or demographic conditions not only broadens students' understanding of Earth, but also provides them with an effective system of conceptual pegs on which to hang future data—geographic, scientific, historical, socioeconomic or cultural.

QUESTION 2: When You Compare Sets of Data Do You Spot Relationships that Occur Regularly?

Setting the stage: Provide the students with atlases (preferably some that show data plotted on several maps rather than just on one), reference books and school texts that can support or spur investigation of patterns of geographic relationships, large chart paper and file cards or loose-leaf notebooks. Then review geographic relationships studied in Chapter 1 or begin a new investigation. Climate, vegetation and physical features offer the most easily recognized and most consistent patterns—although few patterns are without exceptions. Be sure the relationships under investigation can be described in broad enough terms to be applied to many areas, for example, prevailing ocean winds, coastal mountains and heavy precipitation; length of growing season and types of crops; height of dominant vegetation and amount of precipitation.

Comparing relationships and finding patterns: Have each student identify a geographic relationship and list the key characteristics of the relationship. For example, in southeastern Alaska the *above-average precipitation* is caused by *moist, mild air* from the sea carried on *prevailing winds* against the cold *coastal mountains*. (This pattern occurs in other parts of the Pacific Northwest, the west coast of Norway, New Zealand, the northwest coast of Honshu, Japan, and the northeast coast of Hawaii.) Next have teams of students survey maps of the world, or of North America if the pattern occurs frequently, and judging from the maps only, select sites in which the relationship might exist. Once sites are selected, confirming or disproving data should be gathered from textbooks and recorded in a class reference file or loose-leaf atlas. To emphasize the diverse regions in which a pattern occurs, post a world map and have students point out patterns with yarn arrows and a labeled illustration of the kind of relationship that exists.

Raising questions: Focus a discussion on geographic relationships by using some of the following questions: How many key characteristics were necessary to define the sample relationship? (The number will depend on the complexity and preciseness of the relationship.) When searching for other sites in the pattern, which element (e.g., coastal mountains, precipitation amounts or prevailing winds) did you look for first? Why? Do all relationships fall into a pattern? Give examples of some that don't. (Mountains do not always exclude railroad travel as they do in Alaska; mountain areas in Switzerland, Norway and Austria have railroads.) How would plotting the places in which the pattern occurs help you understand the climate (crops, transportation, history, art) of the area?

Varying the setting: A natural outgrowth of discovering one pattern is determining if characteristics opposite from those in the first relationship will also fall into a pattern. Students might contemplate such questions as: Are regions on the landward side of coastal mountains drier and less mild than those on the seaward side? Are dry, hot regions less heavily populated than mild, moderately wet areas? Since the Earth is not a checkerboard and humans are unpredictable, predicting patterns and relationships can never be completely successful. But the range of investigation raised by plotting one set of relationships can be as varied and inventive.

QUESTION 3: When Data Isn't Plotted Can You Guess What Exists?*

Setting the stage: Equip the class with a globe, a large physical map of the central United States and the Canadian shores of the Great

---

*The skill of inferring data requires considerable experience with and knowledge of geographic and demographic conditions. Use this activity with those points in mind.

Lakes, regional or state maps of that area, travel guides of the area, 3-D map plotting objects (clay, flags, etc.) and large paper or poster board. Then refresh students' ability to make "obvious" inferences from plotted data by using a globe. Point out an area on the globe and make a geographic statement about it. Have students decide if the statement is likely to be true and why. (You are, in effect, making the inference for the students; they must be able to discover the relationship(s) that supports the statement.) For example, you might say that Chile (pointed out on the globe) has a large fishing industry. The class would most likely decide your statement is true by noting that nations with long, ice-free coastlines generally have productive fishing industries.

Making inferences from plotted data: On the map of the central U.S., plot the position of early French trading posts and forts (see the illustration on p. 75). Show the name of the fort and the date of settlement but do not specify the purpose of settlement or the origin of the founders. Then post the list of questions, or a few selected questions appropriate to students' abilities, from the "Raising questions" section.

Now have students study the map to find out if, based on experience with geographic and demographic relationships and conditions, they can answer the posted questions with educated guesses or knowledgeable inferences. Encourage the students to use such clues as the sequence of settlement dates, the distance between settlements, the place names that are not French (students may be able to separate Indian names from French). To well-read students, the place name Grand Portage ought to give a clue to the use of canoes, and the position of settlements at straits and other strategic travel spots ought to give a clue about why the site of a settlement was chosen.

When the students have made a substantial number of sound guesses, supply them with detailed maps and travel guides so that they may verify their guesses and identify the trading posts by their present names. (Many names are retained, a few are translations—La Baye to Green Bay—and some have entirely different names.)

Raising questions: Post these questions next to the map and ask the students to answer as many as possible: The names of most of these settlements are given in the language of the white settlers—what country did they come from? When did they come? How did they travel? What makes you think so? In which direction did they expand their settlements? Why do you think the specific settlement locations were chosen? What were the main purposes of the settlements? Do any of these settlements exist today as towns or cities?

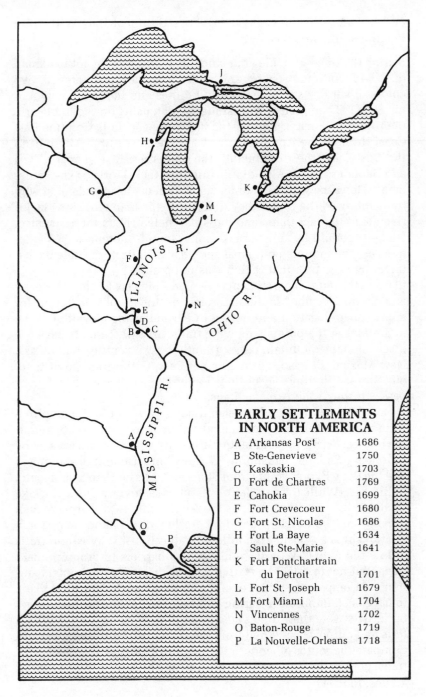

QUESTION 4: If You Collect and Plot Additional Data Will Other Relationships Be Revealed?

Setting the stage: Provide your students with a simple political map of the 48 contiguous states and a physical map for reference, two colors of yarn or string for plotting routes and modeling clay and map tacks. Then plot, or have students plot, using one color of yarn, the routes of some of the most famous pioneer trails from the East Coast through the Appalachians, from the Mississippi-Missouri to the West Coast (these can include the Wilderness Road, Oregon Trail and Santa Fe Trail). Also plot (using map tacks) forts or key settlements along the way. Encourage your students to speculate on why trails followed the routes they did. (Since the map will show major rivers and the fact that some trails paralleled rivers for a distance, students should be able to make some sound inferences. Wide detours around mountainous areas should also indicate to students that travel by wagon in that region was impossible.)

Plotting the data: Have students use modeling clay to plot the Rocky and Sierra Nevada Mountains, referring to the physical map for approximate location and noting in particular the elevation of the mountains at the points where the trails cross. If students have not discovered the mountain passes (South Pass, Wyoming, Raton Pass, New Mexico, El Paso, Texas, Donner Pass, California), point them out. Similar plotting may be done for the Cumberland Gap and other passes through the Appalachians.

Now have students use a second color yarn to plot the approximate routes of the transcontinental railroads (c. 1860–1880) and, if the interest exists, the four main cattle drive routes from Texas to the railroads—Sedalia, Chisholm, Western and Goodnight-Loving—and the Pony Express route from St. Joseph, Missouri, to Sacramento, California. (While the paralleling of routes seems obvious if you take time to think of it, a map brings the relationship into sharper focus.)

Raising questions: Help children further understand geographic relationships by posing the following questions: Why would trails follow the rivers? Why were mountains barriers to group pioneer travel but not to travel by Native Americans, trappers, hunters and miners? Compare the routes of pioneer trails: Which seem most likely to have been impossible during winter? Why? When railroads were built, why were their routes so similar to the pioneer trails? Which forts and outposts benefited from the railroads? From cattle drives? Compare the routes of modern U.S. superhighways to those of early trails, both in the east and the west—do the highways follow the old routes, too? Why?

QUESTION 5: When You Study Relationships Do You Find One That Doesn't Fit a Pattern?

Setting the stage: Provide a physical map and a political map of Western Europe that shows Belgium, the Netherlands, northeastern France and the western edges of West Germany and street maps of your local area. Use the maps of Belgium and its environs to demonstrate one of the classic examples of a relationship that doesn't fit the pattern—at least not at first glance. On the political map, divide Belgium roughly in half with the northernmost provinces of West Flanders, East Flanders, Antwerp and Limburg above the line, Liège, Luxembourg, Namur and Hainault below the line and Brussels and its surrounding province, Barbant, straddling the line. Explain to the class that the dividing line is the approximate boundary of the two languages in the country—Flemish, a language related to Dutch and other Germanic languages, and Walloon, a language related to French and other Latin-based languages. Also explain that language boundaries are usually older than political boundaries and that the former boundaries often follow or were caused by a physical barrier. Have your students locate the Flemish-Walloon line on the physical map to see if they can spot a geographic barrier large enough to enforce a distinct difference in language. (Since the Ardennes region, a reasonable possibility for a geographic barrier, does not come close to following the line, the answer is *no*.) Then explain that up until the fourteenth century a very dense forest, the Carbonnière (source of wood for carbon and charcoal), covered what are now the northern provinces of Belgium. Travel through the forest was limited and it became the logical dividing line between northward-spreading French culture and the westward-spreading Germanic cultures. Today there is no forest, but its earlier impact was so powerful that the tiny country continues to be determinedly bilingual and bicultural.

Reading data and making inferences: Distinct exceptions to a pattern, such as the one just described, are not lying around waiting for beginning map detectives to discover. But it is possible to discover some local physical or cultural exceptions to patterns by using a map, observation skills and a little imagination. Ask students to study a local map or to observe local customs for examples of man-made features that do not fit an "accepted" pattern. Examples might include a small store where, contrary to custom, drivers park parallel to the street rather than at right angles (a pair of gasoline tanks, long removed, once made such parking necessary; the habit persisted). Or a straight highway suddenly makes a sharp bend for no obvious reason (30 years ago the farmer-owner refused to sell his pasture and forced the road builders to detour). The time zone line through the state makes a zig away from the straight line it has been following. Or a state or town boundary line follows the Mississippi faithfully except where tiny loops of land seem to belong to the "wrong" state. Often the real reasons for these exceptions cannot be found in books or maps. Interviews with lifelong residents and local historians or surveys of old photo albums may be necessary. They, in turn, may reveal some additional oddities of local geography, custom and construction.

Caution students that not all exceptions can be eventually proved to fit a pattern. For example, French-speaking Canadians and Portuguese-speaking Brazilians—exceptions to the language of the region—are outgrowths of historical/political events, not geographic barriers. The exceptionally straight U.S.-Canadian border is defined mathematically, not by a natural barrier.

Raising questions: The following questions may help students better understand pattern exceptions: What do you have to know or find out before you find an exception? (You must know the usual pattern, how relationships are usually formed.) Would older maps of the area being studied be helpful in explaining why the exceptions exist? How? If you question local residents for information, how will you choose the people to interview? Is it easier to investigate pattern exceptions in a city or small town? Why? (This depends on the viewpoint and type of investigation. Written records and pictures are often more available in city records and newspaper morgues; but lifelong residents and handed-down stories are generally more easily discovered in small towns. Also, it may help to find reasons for exceptions if students remember that automobiles, road construction equipment and techniques, planes, electricity and modern communication devices are basically twentieth-century conveniences.) Would the condition or elevation of the land make construction difficult without modern machinery? Could the shape or size of land or water features change themselves or be changed by humans? Did the shape or size of the town change in all directions at once or in only one direction? Why?

The intent of this book is to help you convince your students that a map is an exciting device for unlocking the real and imaginary worlds they meet in print or film. Chapters 2, 3 and the first half of this chapter present the kinds of skills children need to use maps fully, and ways students can develop or practice these skills. This last section of Chapter 4 works with four ways maps can challenge students' imagination and actions—as scene setters, as incentives and as clues.

THEME C: Maps Are Keys to Adventure, Exploration, Mystery and the Future

QUESTION 1: If You Had a Map of the Journey Could You Imagine What an Incredible Escape They Made?

Setting the stage: For older students, read aloud some of the episodes found in *Shackleton's Valiant Voyage* by Alfred Lansing, a recounting of Shackleton's incredible escape from his ship trapped in antarctic ice. (For younger students, *The Incredible Journey* by Sheila Burnford or *Brighty of the Grand Canyon* by Marguerite Henry offers travel tales set closer to home.) Provide the class with a large map of Antarctica and/or large drawing paper for a class plotted map.

Tracing the route, sharing the dangers: If you read *Shakleton's Valiant Voyage*, have a group of students construct a large-scale map of the south polar region so that listeners can more fully sense the isolated settings, the slow but desperate progress of the trek, the passage of time and the stranger-than-fiction obstacles to be overcome. (The book has a small but excellent map as an end paper that can be used as a model.) Use of evenly spaced concentric circles to represent parallels—80°S, 70°S, 60°S, 50°S—and evenly spaced spokes to represent meridians makes reproduction of a south polar projection relatively easy and accurate. If you use *The Incredible Journey*, have some students make a Triptik-style (easy-to-read travel) map with a predetermined scale of miles; ask the students to plot the routes the animals followed and the sequence of events as they occurred in the story. The map contained in *Brighty of the Grand Canyon* can be enlarged for class use.

Raising questions: As you discuss the book, try to focus on the following questions: Considering the distance the travelers had to cover, would you have predicted a successful trip? Why? As you traced the route, which obstacles could you anticipate? Which came as surprises? Why? Does plotting each step of a trip on a map make the trip seem more real (exciting, dangerous)? Why? How can "seeing" a trip help you remember what happened during it? If you retraced the route the characters took, can you retell the story? Explain the dangers? Relive the adventures?

Varying the setting: Encourage your students to extend or intensify any study or recreational reading that involves specific settings by referring to maps. Literature (fiction, biography, historical novel, travel journal) and history are filled with many opportunities to use maps as keys to both adventure and understanding. Pooh Corner, Oz, Swallowdale and Treasure Island lose some of their magic without a map. The trials of Johnny Tremaine and Julie of the Wolves, the labors of Paul Bunyan and Ulysses, the adventures of Kim and Tom Sawyer seem to multiply when projected through a map. Historic Santa Fe and Quebec, the shattering "March Through Georgia" and the "Trail of Tears," the forbidding Tower of London and the Great Wall of China, fabled Machu Picchu and Vineland move right into the classroom when seen on a map.

QUESTION 2: Would an Expedition Help Fill in the Empty Spaces on the Map?

Setting the stage: Provide the class with examples of early world maps, explorers' reports or journals. Then post a list of explorers and adventurers appropriate to students' understanding of history (an almanac is a good starting point). Choices might include: Neil Armstrong and Buzz Aldrin; Meriwether Lewis and William Clark; Richard Burton and David Livingston; Fridtjof Nansen, Vitus Bering and Willem Barents; Roald Amundsen, Oliver Perry and Robert Scott; James Cook and John Ross; Henry the Navigator, Lief Erikson and Marco Polo. Or list famous cities or regions that were objects of great searches: the North and South Poles, the Northwest Passage, the ultima Thule, the source of the Nile, the route of the Congo, passage around South America, routes to the Indies, the edge of the world, the Great Southern Continent. Also provide modeling clay or map flags.

Researching and plotting the data: Challenge your students to research and demonstrate how the efforts of a particular individual or group filled in the empty spaces on a map. Many expeditions in the last century, especially those into central Africa and to the poles, were published. Often a librarian can turn up a long-forgotten copy or news clipping of such a trip. Skim or have students skim the explorers' reports or clippings to locate a few key landmarks that appear on a modern map. Then, using these points, plot on a map some of the main features along the route of the trip and compare them to the region as it exists today.

Raising questions: As you compare the old and new maps, pose some of these questions: What kinds of data did the explorer add to the map? What kind of data was he looking for? What did he know about the region he was exploring? What surprised him about the unex-

plored territory? What special tool (e.g., compass, sextant, camera) did he use in exploration? How did the region's position on Earth affect the expedition? Why hadn't the region been mapped earlier? Was any special report made of the expedition? Who was most interested in the report? Why? How has the region changed?

Varying the setting: In addition to expanding our knowledge of land and ocean as they exist today, we can expand our knowledge of where prehistoric and ancient peoples settled and what geographic conditions were like at that time. Archaeologists, paleontologists and geologists build maps based on evidence left by animals and plants. For example, if layers of sea shells are found a thousand miles inland, then there probably was a sea in that place long ago. If bones of musk oxen, who live in arctic climates, were found in Kentucky and other south central states, one might deduce that the climate in that area in the not-too-distant past must have been frigid.

Encourage the students to plot the locations of ancient cities, water bodies, paths of nomads or man-made features in order to learn how scientists and historians "fill in the blank spaces" in maps of the ancient world; such study can also help them to better understand which geographic conditions hindered pre-Industrial human activity and which conditions facilitated it. For example, plotting the known sites of prehistoric man in East Africa points up the existence of a mild climate and a water supply now long gone; plotting the known sites and dates of early man in North America points up paths of migration; and plotting the sites where wampum or coins have been found points up the extent of trade routes.

On the local level, students might try filling in the blank spaces on a town or regional map in a recent, but for youngsters, historic, time, e.g., 1965, before the big flood or tornado. Buildings, streets and other landmarks of the present would appear on the map only if they were in existence in that period. Student archaeologists, in effect, can duplicate on a primitive level the data collection plotting process used to map the ancient world. If local buildings were constructed after the year under study, inferences can be made, and checked out, about what occupied the space earlier. If highways were constructed straight and level in a far from flat landscape, the pre-highway landscape can be guessed and checked against old maps, photos or local residents' memories. Final data may be plotted on a flat map or in three-dimensional form.

QUESTION 3: If You Plot the Facts of a Puzzle, Can You Find Clues About the Missing Pieces?

Setting the stage: Bring to class a copy of Thor Heyerdahl's *Ra II* and/ or *The Bermuda Triangle* by Charles Berlitz, an almanac and a large map of the area described in the book(s). Skim the initial chapters of the book and then outline to the class how the author plotted points on a map to help him resolve a mystery. (The two books should not be considered equally significant. Heyerdahl's research is without question far more scholarly and respected, but both authors do use maps as a tool to study a mystery.) On a map of the appropriate area, plot the evidence (sites where reed boats were found or where planes and boats were lost) that sparked the author's investigation. Help students understand that while individual occurrences of data did not seem related, when plotted on a map a pattern seemed to appear.
Plotting and analyzing data: List a number of mysteries (see the following suggestions) and challenge students to discover missing pieces and probable causes using maps and other research tools. Remind your students that *change* and *exception* often are related to *time*, and that comparison of a region with the same region at another time may be the best route for initial investigations.

Puzzles and Mysteries

- ▶ Some Australian animals are found nowhere else in the world.
- ▶ Although the modern horse is not native to the New World, there are herds of wild horses both in the western United States and on a few East Coast islands.
- ▶ The Loch Ness monster is assumed to be a descendant of seagoing creatures trapped in a fresh-water lake.
- ▶ China is the home of spaghetti.
- ▶ Timbuktu, on the southern edge of the Sahara Desert, was a large city with a famous university and great markets in the fourteenth century.
- ▶ Of the 30 largest cities in the world, only three receive on the average less than 20 inches of precipitation a year—Los Angeles (approx. 13″), Teheran (approx. 9″) and Cairo (approx. 0.1″).
- ▶ The number of newspapers for each 1000 inhabitants ranges from 1766 in the U.S. and 1093 in West Germany to 1 per 1000 in Liberia, Guinea and Senegal.*

Raising questions: As the students attempt to solve the puzzles, have them keep the following questions in mind: What is unusual about the puzzle data given? Where does the data occur? What physical conditions—distance, barriers, climate—are important in the puz-

---

*Based on 1972 figures.

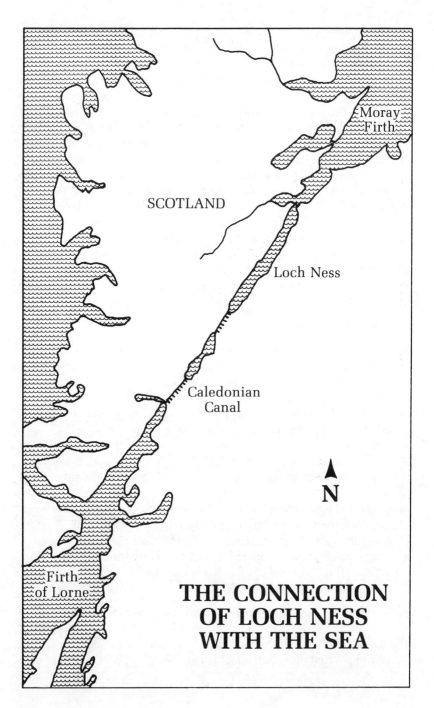

zle? Would time have affected the data? How? Would human activity or invention have affected the data? How? Are the unusual parts of the puzzle likely to change again? Why?

Varying the setting: Students may be interested in collecting and studying treasure maps, such as maps that indicate the location of known sites of Spanish treasure ships or sites where treasure is thought to be (Oak Island, Nova Scotia and wells in Mayan Mexico). Since these maps require little more than the ability to read map symbols, interpretation and prediction skills are not reinforced, but incentives to imagine a treasure-hunting expedition are high. Often the puzzle-resolving problems are not map-related but technological problems, such as raising a ship or drainng a well. However, students might speculate on why such recovery efforts have not been completed.

QUESTION 4: How Far Can You See into the Future Using Maps?

Setting the stage: Provide your students with a collection of maps that show changes of conditions (physical, biological, demographic) over a period of years or review some maps gathered for Question 2 in the second half of Chapter 4. Ask students to point out or emphasize with three-dimensional objects any conditions that show obvious change, for example, steady expansion of a city in one direction could be indicated by arrows or by yarn outlining the increasing city limits; the decline in the wildlife population could be shown with map tacks of various colors, each representing the same quantity but decreasing in total numbers as the yearly total of each species decreases.

Finding patterns, predicting change: Have each student choose a condition that (1) interests him or her and (2) changes enough to be easily detected over a short (in human terms) period. Ask the class to find or make maps that present their chosen conditions today, in the recent past and in the distant past. The range of topics and time limits is vast: daily weather over a week, weekend traffic density in various parts of a city during the past month, the spread of air pollution over a five-year period, the rise of black-controlled governments in Africa over the last 15 years, the change of yearly harvest/precipitation totals in the western Great Plains states over the last 75 years, the growth and/or decline of American railroads and airlines over the last 100 years. Have the students determine what the pattern is and make a map showing the condition in the future. (Be sure students stick to basic patterns for their initial predictions, i.e., discourage attempts to find long-range patterns for hurricanes, earthquakes, droughts.)

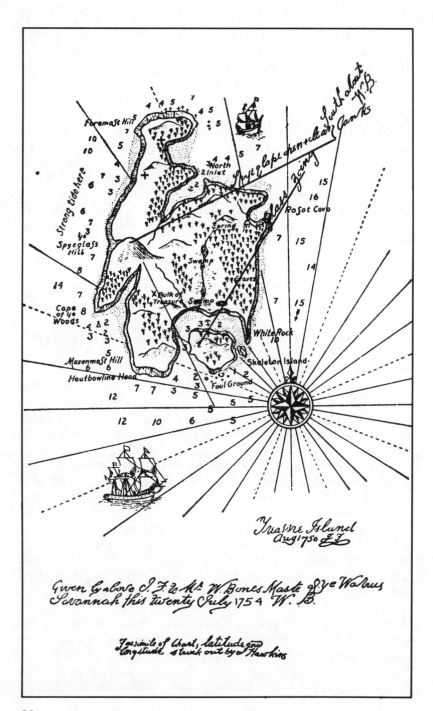

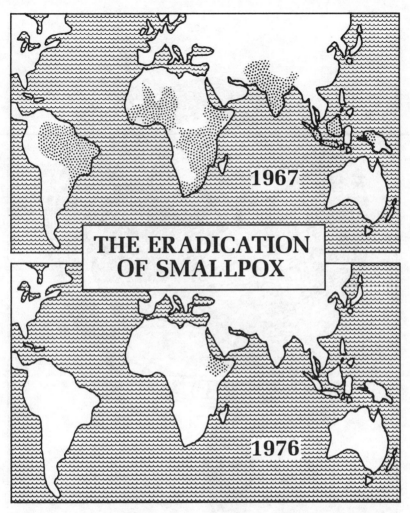

THE ERADICATION OF SMALLPOX

Raising questions: Help students learn to recognize and predict patterns by asking the following questions: How many maps did you need in order to find a pattern? Does the pattern seem to move in only one direction (up, down, westward, toward complete control) or does the pattern seem to follow a cycle (good years, then fair years, then poor years, then good years again)? Do you anticipate anything —geographic or human—halting or changing the pattern (a gas shortage would halt some of the weekend traffic)? Does the pattern come from a negative or positive condition? What makes you think so? How far into the future can you make a reasonable prediction? What makes that period reasonable?

Resources

Asimov, Isaac. *Today and Tomorrow and . . .* Garden City, N.Y.: Doubleday, 1973.

Asimov has compiled several of his articles on time, space, biological rhythm and population in this book. Although not easy reading for the general public, the articles are still fairly much in laymen's terms. Of special interest is a description of the Linnaeus "Flower Clock," which tells the hours of day, from 6 a.m. to 6 p.m., using flowers as timekeepers.

Bennet, E. D., ed. *American Journeys: An Anthology of Travel in the United States.* Convent Station, N.J.: TravelVision® A Division of General Drafting Co., Inc., 1975.

Travelers, travel routes and travel methods, from birch-bark canoes and canal barges and Conestoga wagons to the lunar landing of Eagle, are described briefly, often in words of a traveler from the period. There are notes from Lewis and Clark, comments from an early missionary traveling in Hawaii, accounts by a cattle driver, gold rush participant and dog sled traveler and a few simple maps.

Cameron, Ian. *Lodestone and Evening Star: The Epic Voyages of Discovery 1493 B.C. to 1896 A.D.* New York: E. P. Dutton and Co., 1966 (out of print).

With maps, photographs, diary quotes and prose appropriately flavored with excitement and adventurous spirit, Cameron traces the routes of famous and less well-known seafarers and explorers who rolled back the maps of the world. The voyages begin with an expedition sent out in 1493 B.C. and end with one in 1896. In between are the adventures of seamen from Phoenecia, Rome, Portugal, Spain, Britain, Ireland and Scandinavia, and the sometimes successes, sometimes failures of Pythias, Saint Brandan, Leif Ericsson, Cabot, Barents, Cook, Ross and DeLong.

Collingwood, G. H., and Brist, Warner D., revised and edited by Devereux Butcher. *Knowing Your Trees*. Washington, D.C.: The American Forestry Association, 1974.

In addition to identifying North American trees, their habitats and growth cycles, this book includes many maps of state, regional and national occurrences of various species.

Darwin, Charles P. *Voyage of the Beagle* (abridged by Millicent E. Selsam). New York: Harper & Row, 1959.

Voyage of the Beagle introduces young readers to animal life, prehistoric fossils, flora and geography along both coasts of South America and the Galapagos Islands as they were over a century ago. The maps and time period make the basis for fine map comparison studies.

Hay, John, and Farb, Peter. *Atlantic Shore: Human and Natural History from Long Island to Labrador*. New York: Harper & Row, 1966.

This book maps and describes the various regions from temperate Long Island northward to rocky Maine, the Bay of Fundy and barren Labrador shores. It is a good source of data that could be plotted, e.g., vegetation zones, Gulf Stream and Labrador Current paths, bird and fish migration routes.

Hulbert, Archer. *Historic Highways of America* (16 vols.). New York: AMS Press, Inc., 1902–05.

Indian Thoroughfares (vol. 2) and *Portage Paths* (vol. 7) are perhaps the most interesting of this series, but all are good sources of data about early American travel routes, forts and settlements. Although dated in language, the series should challenge readers who want an in-depth look at early travel in America, written by authors closer to those times.

Lasker, Roy. *Washington in Flashmaps*. Chappaqua, N.Y.: FLASHmaps, Inc., 1975 (paperback).

Washington, D.C. is mapped from over 40 demographic viewpoints—everything from art galleries to zoos, highways to taxi zones, restaurants to bicycle routes. It is a clear, concise, colorful, convenient and inexpensive atlas that can demonstrate use of map keys, north arrows that *aren't* at the top of the page and the organization/selectivity necessary for effective map making and reading.

Minnesota Environmental Sciences Foundation, Inc. *Contour Mapping.* Washington, D.C.: National Wildlife Federation, 1972.

This step-by-step, plot-by-plot booklet introducing the hows and whys of contour mapping is both practical and conceptually sound. The projects suggested might well be used as the culminating projects for map study.

Murphy, Rhoads. *Introduction to Geography.* Chicago: Rand McNally and Company, 1966.

Here is an excellent basic reference for various geographic factors and their implications for human activity. Well illustrated with maps and photographs, this book also contains chapter bibliographies for additional data and a table of world-wide economic and demographic statistics.

Powers, Edward P. *Fair Weather Travel Among Europe's Neighbors.* New York: Mason and Charter, 1975.

This book is designed for tourists visiting the Mediterranean countries of Africa and the Middle East. But the guide can provide students with maps and data about climate, roads, airports, mountains, deserts, population and economic activity in some ancient and far from familiar countries.

Schneider, Stephen H., and Mesiorw, Lynne E. *Genesis Strategy.* New York: Plenum Publishing Corp., 1976.

Here is one of the many new books warning of climate changes that could have far-reaching impact on the world's food supply. Schneider and Mesiorw carry their concern beyond the scientific community to the general public, tracing past climatic changes as well as recent ones. Whether or not your class studies the food crisis, the statistics and examples cited here provide intriguing data for tracing the impact of climate on human activity.

Schwartz, Steve. "Oak Island; Update on the Money Pit." *Yankee* (March 1976).

Sometime between 1695 and 1720, something—treasure, it's hoped—was buried on Oak Island in Mahone Bay, Nova Scotia, and, since 1795, folks have been trying to discover what really is in the Money Pit. Schwartz reports what the status of the investigation is and includes maps and a diagram of the pit.

Stewart, George R. *Names on the Land.* New York: Houghton Mifflin, 1967.

Whether used to demonstrate the effect of culture on place names or as a resource for data to plot on maps, this book is truly fascinating. In it you can learn how early settlers' interpretations of Native American languages distorted the native place names, why some lakes have the word *Lake* first and others second (Lake Superior, Lake Champlain, but Great Salt Lake), the origins of King of Prussia (Penn.) and Truth or Consequences (N.M.) and many other distinctive and not-so-distinctive names for American towns, lakes, mountains, canyons, river crossings and forest clearings.

Taber, Robert W., and Dubach, Harold W. *1001 Questions Answered About the Oceans and Oceanography.* New York: Dodd, Mead, 1972.

The title of this book speaks for itself. The topics range from currents and tides to sea ice, the physical properties of sea water and marine geology—all in relatively easy-to-understand language.

Thrower, Norman J. W. *Maps & Man.* Englewood Cliffs, N.J.: Prentice-Hall, 1972.

Tracing the history of maps and map making, Thrower's book contains many examples of early maps as well as modern ones. Of special interest is an 1855 dot map showing cholera deaths in London (an early use of maps was solving medical mysteries). The appendices include a list of major map projections, their characteristics (including distortion) and principle uses and a glossary of mapping terms.

Von Frisch, Otto. *Animal Migrations.* New York: McGraw-Hill, 1969.

Von Frisch focuses on migration of all types of animals—birds, mammals, butterflies—in various regions of the world. Well illustrated with photographs and maps, the books is a good source of information for concept building and data for use in mapping.

Sources of Inexpensive and/or Special Maps

Geological Maps of the Moon. Topographic Maps, Silent Guides for

Outdoorsmen. Branch of Distribution, U.S. Geological Survey, 1200 South Eads St., Arlington, Va. 22202.

For further information: U.S. Geological Survey, Map Sales Office, Sunrise Valley Dr., Reston, Va. 22202

Volcanoes. U.S. Department of the Interior and the U.S. Geological Survey both print pamphlets under this title. Both are available from the Superintendent of Documents, U.S. Government Printing Office, Washington, D.C. 20402.

Descriptive List of Treasure Maps and Charts, compiled by Richard S. Ladd, 1964. Superintendent of Documents, U.S. Government Printing Office, Washington, D.C. 20402.

Wreck Information List (coastal waters of U.S.), published by the U.S. Hydrographic Office, 1945–46. Available at Photo Duplication Service, The Library of Congress, Washington, D.C. 20540.